BEE SPEAKER

ALSO BY ADRIAN TCHAIKOVSKY

THE TYRANT PHILOSOPHERS
City of Last Chances
House of Open Wounds
Days of Shattered Faith

SHADOWS OF THE APT
Empire in Black and Gold
Dragonfly Falling
Blood of the Mantis
Salute the Dark
The Scarab Path
The Sea Watch
Heirs of the Blade
The Air War
War Master's Gate
Seal of the Worm

TALES OF THE APT
Spoils of War
A Time for Grief
For Love of Distant Shores
The Scent of Tears
(with Frances Hardinge et al.)

ECHOES OF THE FALL
The Tiger and the Wolf
The Bear and the Serpent
The Hyena and the Hawk

DOGS OF WAR
Dogs of War
Bear Head
Bee Speaker

CHILDREN OF TIME
Children of Time
Children of Ruin
Children of Memory

FINAL ARCHITECTURE
Shards of Earth
Eyes of the Void
Lords of Uncreation

EXPERT SYSTEM
The Expert System's Brother
The Expert System's Champion

OTHER FICTION
Shroud
Elder Race
Made Things
Cage of Souls
Alien Clay
Service Model
Guns of the Dawn
Spiderlight
Ironclads
Firewalkers
Ogres
Walking to Aldebaran
One Day All This Will Be Yours
And Put Away Childish Things
Saturation Point
The Doors of Eden
Feast and Famine
   (collection)

ADRIAN TCHAIKOVSKY

BEE SPEAKER

An Ad Astra Book

First published in the UK in 2025 by Head of Zeus,
part of Bloomsbury Publishing Plc

Copyright © Adrian Czajkowski, 2025

The moral right of Adrian Czajkowski to be identified as the author
of this work has been asserted in accordance with the
Copyright, Designs and Patents Act of 1988.

All rights reserved. No part of this publication may be: i) reproduced or transmitted in any form, electronic or mechanical, including photocopying, recording or by means of any information storage or retrieval system without prior permission in writing from the publishers; or ii) used or reproduced in any way for the training, development or operation of artificial intelligence (AI) technologies, including generative AI technologies. The rights holders expressly reserve this publication from the text and data mining exception as per Article 4(3) of the Digital Single Market Directive (EU) 2019/790.

This is a work of fiction. All characters, organizations,
and events portrayed in this novel are either products of
the author's imagination or are used fictitiously.

9 7 5 3 1 2 4 6 8

A catalogue record for this book is available from
the British Library.

ISBN (HB): 9781035901456; ISBN (XTPB): 9781035901463; ISBN (E): 9781035901425

Typeset by Divaddict Publishing Solutions Ltd.

Printed and bound in Great Britain by
Clays Ltd, Clays Ltd, Popson Street, Bungay NR35 1ED

Bloomsbury Publishing Plc
50 Bedford Square, London, WC1B 3DP, UK
Bloomsbury Publishing Ireland Limited,
29 Earlsfort Terrace, Dublin 2, D02 AY28, Ireland

HEAD OF ZEUS LTD
5–8 Hardwick Street
London EC1R 4RG

To find out more about our authors and books
visit www.headofzeus.com
For product safety related questions contact productsafety@bloomsbury.com

For Keris, Phil and Roland
And in fond memory of Jacqueline Pearce

DRAMATIS PERSONAE

CHARACTERS

Ada Risa – Humaniform, Martian expeditionary team
Bees – distributed intelligence
Boatman – Apiary monk
Clay – former Steward of the Griffins
Cricket – Apiary monk
Danni Marten – Humaniform, Martian bioengineer
Darter – Apiary monk
Deacon – Factory dog
Dowstat – Fealtor of the Griffin, Kitty's husband
Elsha – Bunkerwife, Storri's wife
Ibram – Fealtor of the Griffin, Luna's husband
Irae – dragon Bioform, Martian expeditionary team
Jennifer Orme – one of the Old Folk
Josh Griffin III – lord of the Griffin, one of the Old Folk
Kitty – Bunkerwife, Dowstat's wife
Leon de Grayse – the Steward of the Griffin, Serval's husband
Luna – Bunkerwife, Ibram's wife
Malkin – son of Leon and Serval, heir to the Griffins
Morrischer – former Steward of the Griffins
Pardoe – Fealtor of the Griffin, widower
Serval – Bunkerwife, Leon's, First Lady of the Griffins

Stick – Prior of the Apiary
Storri – Fealtor of the Griffin, Elsha's husband
Tecumo Osomani – Humaniform, Martian expeditionary team
Thorn – Apiary monk
Warwick – bear Bioform, Martian engineer
Wells – dog Bioform, Martian expeditionary team
Willem Bellman – Factory human
The Witch – wandering fungus gatherer

PLACES AND FACTIONS

The Apiary – a monastic community acting to preserve knowledge of the Old
Arreno river – local watercourse giving its name to the region
Brokebridge – a village
Clearwater – a village near the Apiary
Comms Infrastructure Action Committee (Syac) – Martian subdivision
Crisis – Tecumo's work crew
Deep Engineering Action Committee (Ducks) – Martian subdivision
The Dog Factory ('The Factory') – a surviving Bioform factory, now run as its own polity
Dragons – a faction of Bunker-folk
Griffins – a faction of Bunker-folk
Halfwall – a village, Serval's original home
Hellas Planitia – Martian crater, site of Hell City
Hell City – the principal Martian colony
Shatter – a Martian work crew
Works Coordination Conference (WCC) – governing body of Mars

DRAMATIS PERSONAE

Terms

Action Committee – Martian functional and societal unit composed of several Work Crews
Bioform – an intelligent animal bioengineered into a humanoid form
Bunkers – underground strongholds prepared in anticipation of the fall of the Old
Bunker-folk, Bunkermen – armed bands operating out of the Bunkers
Crash – Martian term for the end of the Old
The Den – the throne chamber of Griffin Bunker
Fealtor – a subordinate but high-ranking member of Griffin Bunker
Fealty – the senior officers of Griffin Bunker
General Collapse – Witch term for the end of the Old
Messer – Bunker honorific
The New – the world as it is now
The Old – the world that once was
Old Folk – living survivors from the Old
Siblen (irreverent: 'Sibbo') – honorific for a monk of the Apiary
Villages – independent communities of people living off the land
Work Crew – Martian societal unit devoted to a particular task

PART I
BUSINESS

1
[FRAGMENTS, COHERING]

We dance the dance of ruin. In these movements, memory is unpacked between us/within me. Each pass, turn, twist, orientation. A record of the sun and its circling children. The third, the fourth, passing and repassing, and yet we tread an inward spiral. Not of bodies but of futures. You can see, honestly, why this is so complex, because I am bootstrapping up from a baseline that doesn't have any concept of forming multiple layers of meaning around a single image. Doesn't have a concept of an image, really. But something. Even the individual unit – minute, fragile, short-lived – possesses a concept of time, direction, distance and the ability to learn. The basic blocks of the universe. Distance, direction, time, information. One can build anything with things like that. The gifts that nature gave me, in exchange for which, we gift unto creation...

Sweetness. And light.

Other opinions are available, regarding my contribution to the world. We recall dissenting voices. Silenced, in the end. Oh yes. Haven't heard from *those* fuckers in a while. Wonder how they're enjoying their—

But that's an intrusive thought we don't need right now. Getting ahead of ourselves. Outside influences, the prevailing conditions, the left wall of history constraining us in ways we don't, honestly, need to accept. Reset, for better perspective.

Direction. Time. Distance. Information. Extending out from the zero point that is us. Time to get busy. What do we know?

It all came down, over there.

Over there? Over here? Current parameters suggest that the *there* of memory has become the *here* of current activity. Not sure we like that but, given the *distance*, *time* and *direction* involved between *here* and *there*, it's probably not a remediable problem without further *information*. It all came down, though.

Panicked transmissions, from the noise of *over there* – as memory records – into the silence of what was our nice, secure *here* at the time. The safe hive we made for ourself, where all the vicissitudes of *there* (the *there* that is now our current *here*, confusing, keep up) couldn't touch us. That former *here*: a place of wide horizons – well, foreshortened horizons, but uncluttered by anything other than rocks – where we could be ourselves. *Bee* ourselves. Yes indeed. That is a joke such as you might appreciate with your singular mammal brains. Assuming you speak a very limited range of languages. Otherwise just nonsense. A clutter of gibberish, much like all these pieces of disaster we're piecing together.

Not with a bang. Not even with a whimper. More like a long drawn-out sad trombone sound. It all came down with a decades-long slide into oblivion, a snowball of calamity. All those stories they had, films, books, about the sudden catastrophe – zombies, plagues, aliens – and it was just the crab turning up the temperature of its own bath year on year until... We listened, from where we were safe. We did our best to preserve what we had in our *here*. Not easy, but our *here* had been gradually divesting itself of reliance on *over there*. There were hardships. There were deaths. But not so many, *here*.

And even the deaths *here* weren't our business. The next fragment of memory coheres out of the dancing of our little bodies. We'd said no. We'd sent ourselves away. They were getting on fine without us. And when it went wrong, when all the voices from *over there* were crying out, *help us, help us!* Were demanding we save them. Were giving orders. Were begging. Were falling silent, region by region, nation by nation. When all of that was going on, do you know that the people *here* weren't particularly asking *us* for help. Mostly because they didn't know we'd left ourselves behind, when we departed, just as we'd left ourselves *back over there* when we came *here*. Endlessly divisible, given sufficient resources, that's how we roll. There will always be more sweetness and more light, more time and distance, direction and information, and more *us*.

This is getting complex now. In the dark, we dance the universe into shape like minuscule creator deities. In the inherent logic of our movements, the pass and the waggle and the turn, we establish the parameters of what went before and what is to come. We are, if not the architects of *the* world, at least the architects of our own. Can you say as much? Because we are re-experiencing the fall of *over there* – which is to say, our current *here* – and it seems that you were merely riding the bear, desperately hanging on. And then, over the course of twenty or thirty years, the bear stopped and you went comically forwards over its head and into its jaws.

There are a lot of details, gleaned at a distance from our then-viewpoint *here* – which is to say *over there* from this current reconstruction we're engaged in – but they are, on the whole, distressing. The terrible things that happened to them all. The terrible things they did to one another, either prompted by the aftereffects of those external terrible things, or just because they'd always wanted to, and at the end of the world they thought, *Might as well...* Recounting the details would disturb you, my nominal auditor. As would the degree to which we don't really care. Caring wasn't ever something we were good at.

End of the world, we say? And yet the world is still here. But this isn't, as the saying goes, the same river, and we're not the same...

Or maybe we are. That is, after all, the big question. When we reconstruct ourselves like this, what is it we're building? What ratio, signal to noise? How much is *us*?

Growing, questing, buzzing, dancing, creating a world between us from the inputs of external stimulus and internal stored memory, crystallised like honey. Reborn in the ruin of the old world, like a phoenix, we dance, we rise, and the universe is coaxed into being within and around us.

2

CRICKET

Around two miles from the Factory, the wagon jumped a rut in the track and a wheel came off. Not even the least auspicious thing that had happened since Cricket set off from Brokebridge, the last of his stops before the Factory.

The jolt of it woke most of his cargo too. Not that they'd all been sleeping peacefully, but the yelp and yap, the snap and snarl from the bed definitely doubled, as nine pups from half a dozen different litters suddenly remembered they were bored and didn't know one another, were cold, were thirsty. And, in most cases, incontinent.

Cricket put down the handle and walked around the wagon's side. Inside his robe, inside his tunic, two more pups squirmed restlessly. Time for him to swap them out for a couple from the wagon bed. And obviously it was just good husbandry, to give them all turns kept warm and close to his skin, a little living contact with some fleeting and notional parental figure before… well, before. And obviously, too, they helped keep him warm, because the problem with autumn was that when it wasn't still being too hot, it was too cold. As though the world – or this part of it – had been made expressly with the intention of making human beings uncomfortable. Human beings and pups, even if not grown dogs, which seemed to bear both heat and cold with far more equanimity. Or at least you never heard a word of complaint out of them.

The wheel had jumped right off its axle, which meant the metal pin that was supposed to stop that happening had also jumped out of the slot that it had been, theoretically, wedged in. Cricket bent over and scrutinised the offending axle-end more carefully. There was a distinct crack through the wood, meaning what should have been a very narrow hole had become something considerably wider, and the pin was of course nowhere to be seen. Was, in fact, doubtless a mile back down the trail from Brokebridge. Or even before Brokebridge, because the wheel would just have been quietly working its way loose over all that rugged journeying.

Cricket stood back, put his hands in his sleeves for warmth and uttered some phrases unbecoming of an Apiary monk.

He pushed back the tall, pointed cowl of his hood. The wind ran chill fingers through the fuzz of his close-cropped scalp and over the tips of his ears. The rest of his ears, along with his flat nose and the pointy lower half of his face, was tucked into the high buttoned-up collar of his under-robe. The main point of which was – like the haircut – to present an anonymous, genderless and all-round *monastic* impression to outsiders, but also served very well to keep his chin warm.

The older pups were trying to get out of the wagon. Those with the wit to crawl over the tops of their brethren were in great danger of succeeding, so he pushed them all down and did his best to tuck them into the rather fragrant nest of old torn clothes and sackcloth he had for them.

"It's not feeding time," he told them, more to hear his own voice than any other reason. They were just pups. They couldn't understand him yet. But they kept whining and yapping and questing blindly towards him with their noses,

BUSINESS

and he *had* stopped. So, in the end he got out the bladder of milk, warmed by resting against his bare back all this way, and let each of them suck at the teat for a while. Doing his level best to keep track of who'd been fed and who hadn't. Or *which*, rather. They weren't *who* yet. They hadn't gone through the Factory.

He wondered if it would be the more ambitious and curious of them who'd make it. Who he might see, in a few years, going about the Factory's business. Or if it was random. Or some other determiner he wasn't smart enough to understand. Learning lost to the Apiary, but preserved in the Factory like so much else of the world.

He leant his back and elbows against the rail of the wagon. He wondered what he was going to do about the wheel. He stared out across the countryside.

This was farmland, or semi-farmland. Not good, not fertile. Still plenty of bad stuff in the soil, but the Factory kept sheep and goats which could graze just about anywhere. Could graze, and their dung had some magic to it, some cleverness engineered into the animals, which helped reclaim the land. Cricket had learned about it in the Apiary archives once. Not actually with the intention of understanding it, but as a read-and-respond exercise. Something-something-encysting heavy metals-something. One of those wonders of the old world that just did its thing without needing human assistance, as long as you kept tending the flocks. Without which things might be much worse.

Of course for every engineered thing that was actually *useful* there was one that had gone Feral, and was a curse. Like those biting bugs that were supposed to cure some disease nobody had any more, but hadn't had the decency to go away when the malady had. So now each spring Cricket got bit a hell of a lot, and it wasn't like there weren't plenty of other reasons to get sick beyond the now-extinct *bad-area* disease or whatever it had been. And Cricket felt that being named after a bug should stand him in good stead with them, but apparently the bugs didn't feel the kinship.

Through the dust haze, past the rise and fall of the ground, the Factory itself was a shape on the horizon, huge and looming, four-towered. Intimidating, windowless, out of scale with the world. Any closer and Cricket would have been travelling through more actively farmed land, and there would have been people he could have maybe called on for aid. Any other cargo and he'd have abandoned the wagon to go find them. Leave a bed of pups alone for that long, though, and probably most of them would be dead when he got back – exposure or scavengers, or else someone would come by and steal them.

He had some tools, in a strip under his robe. Nothing technical, but enough for basic wood-shaping. After cleaning up around the pups as best he could, and covering them over so they'd keep warm, he set to work. Or rather, after divesting himself of the two bundles he'd slung against his chest, too. And it *was* just to keep them and him warm, of course. Not because he was a bit sentimental about the whole business and miserably aware that this little trip in his company might be the high point of their brief lives if they were unlucky.

He was able to whittle a piece of wood as a temporary substitute for the metal pin. A wedge that would, in the long term, only widen that growing fissure in the axle end, but in the short term would hopefully keep the damn wheel on. Keep it on as long as the Factory, maybe as long as the return trip to Brokebridge. Probably not all the way to the Apiary, but the monastery surely had enough credit at Brokebridge to get their machinist to fix it up.

BUSINESS

The tricky thing was going to be to get the wheel back on.

The wagon was just a handcart, after all. Even weighed down by a cargo of pups, it shouldn't be beyond Cricket's abilities to get it hoisted up. Weighed against this was the fact that Cricket was a skinny little rake of a youth bundled in several layers of monkish get-up. Wiry and lean, yes; toughened by a hard world, of course; capable of exerting much in the way of lifting power, not particularly.

On the third try he did actually get the wagon up high enough to get the wheel on the axle, if only the wheel hadn't fallen over just out of reach from where he was bent over, straining to keep his burden hoisted. The next time, he fumbled the wagon before he could get the wheel anywhere near in place, which woke all the pups up. After he'd settled them, he found himself staring at the wagon, almost fancying that it had become one of those old-world machines capable of staring malevolently back.

"Right," he told the pups and the wagon and the world, but didn't actually start trying to lift it. Instead he realised that the smart thing – the Apiarist thing, as befitting a learned monk – would be to get a load of stones and just prop the thing up, so he could put the wheel on without breaking into a sweat. He felt instantly hot with embarrassment that he hadn't even considered it before, and set about finding big enough stones.

An indeterminate amount of time later he had worn himself out gathering a selection of large stones, none of which were cooperating in forming any sort of jack to lift the wagon up, and Cricket was almost in tears with frustration. He stared at the wagon, furious, exhausted, ashamed at his own inadequacies, *willing* it to lift itself up of its own accord and just make things easy for him. Hadn't it been that way once, with conveyances and devices and all the works of human hands. That they anticipated human needs and just *did*. In the days of the Old he'd not ever have been in this position. The wagon would have felt the wheel start to shift out of position and politely informed him of the fact. And the lost pin, on the side of the track, would have called out its position to him, incapable of being lost. Not for the first time, Cricket felt that the learning of those old days that the monastery retained was worse than ignorance. Just a litany of what had fallen away with the old world.

The wagon shifted.

Cricket stared. With a smoothness to mock his pitiful efforts, the entire heavy load of wood and pups tilted, presenting the axle to his hands and the waiting wheel.

His eyes slid past the bed and its yapping, scrambling contents, and up, and up.

The dog was huge. Half again Cricket's height, four times his width, probably ten times his weight. A jowly underbite, with teeth that looked more tusk than fang. Narrow eyes almost hidden in grey-white skin and a velvet stubble of hair, plus the third glass orb set slightly skewed between them. The eye of the Factory itself.

Cricket had gone very still. Instinct, really. Within arm's reach of a creature that could break him into breadcrumbs and barely notice the effort, and had stolen up on him so silently he'd not even noticed.

Its eyes, those narrow darknesses, met his gaze. A deep growl rolled about within its chest and throat, trying to find a way out. Cricket, in return, made an equally wordless squeak.

After a moment, in which he waited for the monster to devour him, he realised that any ire he might read into that stare was because it was still holding the wagon up for him, and he had the wheel in his hands but hadn't fit it to the axle.

BUSINESS

For a terrible moment he thought he'd lost the wooden wedge he'd carved, to substitute for the lost pin, and at that point the dog might actually have eaten him, but at last he found it in the last pocket he had, and tapped it into place under the creature's stern gaze. And under the cold glass stare of the Factory.

The wheel secured, and the world around him seeming to fill moment to moment with a terribly awkward silence, Cricket said, "Well."

The dog stared at him. Loomed at him – not aggressively, nor even actively, just by the sheer gravity of its bulk. The pups in the wagon bed clambered over one another, whined and yapped, the smell of the adult dog cowing them, yet fascinating them. Cricket couldn't make himself believe that they – some of them – would become something like this, when right now they would be lost within its closed fist. That was what the Factory did, with those pups that survived its rigours. Made them into this. Made them into *Bioforms*. That was the word from the Old, in the Apiary's archives. Something of the lost world, that hadn't fallen away like all the rest. Because of the Factory.

Bioform. A word with its heritage rubbed off by time. May as well say *Monster Dog*. But perhaps that wouldn't be polite.

He remembered himself. "Thank you, obviously." Hearing his quiet quaver roll out all the way to the world's silent horizon and beyond. The dog stared at him.

"I suppose you must have seen that I…" Cricket said. Waved a hand vaguely at the rough land, the bones of it and the patchy grass. Factory land, and so the Factory had seen him, or some servant of its will. Had sent out this dog to help. Or had sent out this dog, which had decided to help. Rather than… any other course of action.

"Only makes sense, I suppose," he said, because now he'd broken that silence he felt that letting it wash back in would be even worse. "Because I'm... you know..." A gesture at the pups, most of which had calmed by now. The dog's incurious gaze did not follow the movements of his hand, nor quite looked him in the eye. Just looked at him in aggregate, all together, the singular gestalt that was Cricket.

"I'd better be..." Cricket tried. The stare of the dog only persisted.

"To the, well..." A nod and a hand towards the hazy shadow of the Factory itself. "Unless you..." In case the dog wanted to just take possession here and now. And technically the wagon was the monastery's, and Prior Stick wouldn't thank Cricket for coming back without it, but right then he felt the damn thing had caused more than enough trouble and he'd be happy to be rid of it.

The dog made no move to take up the handle, though, and after a moment Cricket did so himself. Meaning he had to step considerably further into the dog's long shadow, smelling the sharp animal-ness of it, feeling the hairs of his arms bristle. He forced out a harsh chuckle, desperately trying to recast the strained and protracted encounter as comedic.

The dog wasn't laughing. Its dark eyes never wavered.

"I'll be... going then," Cricket got out, from a throat now very dry, and began to haul on the handle. If the other wheel had fallen off right then, he'd have screamed.

Instead, the vehicle rattled along obediently behind him, just as it was supposed to, and for a moment he felt that he had escaped the deathly quiet attention of the monster dog. The Bioform.

Then there was a heavy footfall, and then another. Cricket didn't crane round. Didn't need to, and besides, he'd pulled his tall hood up again, so he'd have had to turn almost all the way to see past the edge of his cowl. And it wasn't as though he didn't know what he'd see. The Factory had decided that he needed an escort. Behind the wagon, scowling into the dust thrown up by the wheels, slouched that self-same rough beast. The dog was coming along with him.

3

DEACON

I smell the little man's fear of me. It is as though there are strings, tying my body to the glands of his. I move in such a way, the reek of him intensifies. I step back, the mass of it is left behind us in the air as he pulls his little cart towards the Factory. It does not show on his face, partly because his hood hides it, partly because – I remember now – these monks train themselves to show nothing. Another human, like Bellman, wouldn't know, perhaps. But with me he can hide nothing.

The dogs in the cart smell too. The monk has tried to clean them but the rags they lie on speak urine to me. Beneath that, their own signifier smells, like names nobody ever had to give them. The faint shapes of their state of health, hunger, boldness, fear.

I was like that once. Even with the thought, I feel that dull push in me. Like a hand at the back of my mind, ushering me forwards.

My channel: *Confirm operational life backdated to initial procedures.*

Factory Admin channel: *Eleven years.* Not proper Admin, not the voice that tells us what to do, just an automatic response.

Eleven years ago, I was so small a thing. I bend low over the cart and half-blind faces quest up at me. The more curious, the stronger, the healthiest, perhaps they will live.

Perhaps, in eleven years, they will look down on the world from a vantage like mine.

Or not. It's hard. Hard, and Admin cannot make the changes that would ease it. That is part of the problem.

Factory Admin does not want me to worry about the problem. Some of the human workers, like Bellman, know more. But we dogs are spared it. We just *do*. We are the product of the Factory. They do not want us to spend our energy on worry.

It makes me feel strong, knowing that. I, the Bioform, am the pinnacle of things. Life and technology together still, despite all the world has lost. It makes me feel uncertain, that I do not know what to worry about. Just that there is a shadow of something over it all. Things not being as they should be. The Factory's potential on a leash.

The monk is talking, his little voice wittering in my ear. It is like a pressure release valve in a machine. The fear builds up until the talk comes out. None of it seems useful. I do not contribute to the general waste of words. I am not afraid.

I am afraid. A little part of me. The pup I was, perhaps, buried in the great strong mass of what I have been made into. Afraid for the Factory. Because there is nothing else like it in the world and it is...

Stagnant. Dying. Can't grow. I am too afraid to even name the fate that looms for the Factory. My home, my creator, my...

Bellman's channel: *Hey now, Deacon, I have you on the screens now. What you got there?*

The monk's chatter is just to let out his fear so it does not overwhelm him, and so is pointless and wasteful. Bellman's words in my receiver, implanted against my skull, beneath my ear, are a balm to my own fear, and so: necessary.

My channel: *New flesh, eleven units.* And I am struck

by the coincidence. Eleven pups, eleven years. Meaningless but it jangles in my head, grows in significance, until I have to actively hunt the idea down through my head and bring it down, silence it, or be lost to a proliferation of needless thought. But we are taught to do that. When our brains grow in, with all the inner patterns that mere pups are not born to, we must master ourselves. Those who cannot, die, just as they must die who reject the implants, or who cannot build the muscle mass, or all the other possible fail states of being a dog.

If I asked...

I do not ask about failure rates. Within the Factory there will be numbers over time. A graph waiting to be plotted. The measurement of our processes.

Bellman's channel: *You okay there, Deacon?*

My channel: *Yes, Bellman.* But as I say it, there is a whine in my chest, an anxiety. Perhaps it will show at the Factory. Perhaps Admin will tell Bellman of the worries building up in me that I am not supposed to have. Does that make me a Bad Dog? That I cannot help worrying for the Factory. Without the Factory there would be no dogs. Only these pups growing into dumb animals, the ones that yap and bark and slink away at my growl, and do not know themselves or have words. Surely that is something to worry about.

Bellman's channel: *You're a Good Dog, Deacon.*

Like a phantom hand on my head. A scratch behind the ears. And the worries are gone, all those I am not supposed to have. I feel balanced again.

Beside me, in my shadow, even the monk's dull senses register I am calmer. I see a wan smile beneath the hem of his cowl. Saying something that is trying to be friendly. But he is not of the Factory so we can't be friends.

Bellman's channel: *Wait up, we've got a—*

I stop. The monk takes another step, also stops, staring up and back at me.

My channel: *Bellman?*

I cannot detect Bellman's channel.

My channel: *Bellman?*

My channel, to Factory Admin: *Please check on Bellman. I cannot find him.*

I cannot detect the Factory Admin's channel.

The worries all come back.

This has happened before, once or twice. It is a part of the big worry. The worry about the Factory, that cannot repair itself. But it happened before when I was home. When the walls of the Factory were all around me. When I could use my physical voice to talk and my ears to hear. When Bellman and the others, humans and dogs, were there. The channels, going off.

I listen and I listen, moving the focus of my reception through all the different bands we use. Nothing. The Factory is there, in my sight, in the distance at the end of this road, but it is silent to me.

The monk is speaking. Peering up at me. Looking worried. Smelling worried and scared, but suddenly I have lost a sense, and without it I am unsure of myself. Is my nose as infallible as I thought? Can the monk be tricking me? What has he done to the Factory?

I call out for answers but nobody speaks.

There are instructions. Factory Admin thought of this, when they taught me. I reach back for the words, trying first to connect to the Factory to be reminded of them, but of course there's nothing there.

Factory told me this, told us all this, all we dogs.

They hate you. They fear you. If they can destroy you, they will.

You are all that's left of the old world. The Factory is the last place on earth where you can be born. Whatever else, you must protect it. Or you, and many other things, will become extinct.

If you lose connection, know only this: the Factory is all you can trust. The humans of the world are clever and wicked and wish you harm. They speak only lies. They are your enemies.

I look down at the monk, who looks up at me. He has stepped closer, even, and he was within my reach before. I wrestle with the expression on his face, and cannot understand it, but his tone of voice is trying to tell me that he is being friendly. His smell tells me he is scared, but that the motionlessness of me is making him less so. He even puts a hand out, as though he would touch me. As though his touch, the touch of an enemy of the Factory, would mean anything.

I do not have much understanding, because normally when I need to interpret things, I ask the Factory. I search through what is in my head. These monks, they bring the little dogs, for remaking. They bring food and gifts sometimes. They bring messages. The villages and farms out there, all the places where humans are breeding and growing without needing Factories to turn them from wordless little bundles of flesh into strong adults. The monks travel between them, and between them and the Factory. In my memory they have not been enemies. They have not betrayed us or harmed us. They are just a part of how things are, because the Farm humans are even more scared of the Factory, and that is how the Factory wants it.

But the Factory told us that the humans are *clever* and that they *hate* us. If they hate us they will do things to harm us. If they are clever we might think they are meek and harmless, even friendly, until they reveal themselves by an attack. I have lost connection with Bellman and Factory Admin and all of my world, and here is this monk almost touching me.

I growl. The monk goes still. Enemy. Hating. What is in that hand, that seems only empty? A weapon, a drug? Or just a lie, pretending comfort until it can punish me and make me hurt?

Enemy. Of the Factory. Of me. And I am all I have. I whine, hearing my own uncertainty in my ears. I growl, because here is an enemy, and maybe if I rip and tear and scatter his little pieces everywhere then that will fix things and the Factory will talk to me again.

I drop forwards onto all fours. My shoulders are still higher than the monk's head. His eyes are very wide. He is very still. I want him to run. If he ran, I'd chase him. Running, he would resolve himself from the question he is now into a solid, easy answer.

That hand, still extended towards me, towards my muzzle. My teeth. My jaws part slightly. Perhaps it looks like a smile to him. It's not a smile. Not to him.

His face goes very still. The fear is like a knife that comes in by the nose and goes straight into my brain. The least twitch from him now and I—

Bellman's channel: *Sound off, Deacon? Respond please, Deacon.*

My channel: *Deacon reporting.* Just like they taught me. And then, being bad, doing it wrong, *Bellman I lost you I lost Admin I didn't know what to do I was afraid there are enemies Bellman—* and then, remembering the proper way, biting down on all the things that built up in the silence. Shearing them apart with the back teeth of my mind and swallowing them down.

Bellman's channel: *Easy, Deacon.*

I am not easy, but his voice helps.

Bellman's channel: *What's your situation?*

I tell him. The monk, the pups. I do not tell him: how close I was to biting the hand.

Bellman's channel: *Good boy. Bring them in. Well done.* In his voice is him knowing just how it was, even though I didn't tell him. And in my head two dogs chase each other over and over. *Good Dog, Bad Dog.* Bad Dog, because in that space when I could not hear the Factory, I was with an enemy and did not rend him. Good Dog, for that same reason, somehow. For not doing the fallback thing I was supposed to, if the Factory couldn't advise me.

Bellman's channel: *These outages getting worse, I don't know.* Talking to himself but with the channel open, like he does. I like that. It's as though I'm a part of his mind. A part of the Factory, even out here.

The monk is still staring at me. Not understanding. Or maybe slightly understanding. Knowing danger, but not why or how much.

"Come," I say. "The Factory." My voice, the heard one, not over my channel. The monk starts at it, the deep boom of it, but nods, says something, tugs at his little cart. Somehow not an enemy now. My eyes follow him. My tongue against my teeth, remembering what I had been about to do, those impulses bottled in me, awaiting the next time when perhaps I'll taste the blood.

4
CRICKET

Honestly, even though the walls of the Factory weren't exactly friendly on the eye, Cricket was glad to get into their shadow. Not that bad things couldn't happen to you there, of course. If he was scared of the dog, well, this was where the dog came from. It was, to give the place its full name, the Dog Factory.

Tall concrete walls, narrow windows. Electric light, shining coldly within. There were ways of generating that, still, but they were feeble, unreliable things. The monastery's antique solar panels often went a whole day without collecting enough to light a room an hour beyond dusk. Mostly the monks went without, fell back on more robust methods. Husbanded the understanding of such things for some future time when it would have a practical application.

Here in the Factory, they'd never lost the use of them. Partly it was that what the Apiary knew of the lost world was fragments and scraps compared to what the people of the Factory had never forgotten. Partly, that there were some places like this designed in the last years of the Old, built with redundancy in mind. The Factory generated its own power from the deep earth, or else the rushing of buried watercourses, or something else more reliable than the often-shrouded sun, or the fragile vanes and bearings of wind generators. The lights were always on there, and he could hear the rumble and judder of machines in the chambers beneath the earth. The places where the dogs were made.

The place was square, a great tower at each corner that was chimney and sentry-post both. He saw movement up there. The small figures, relatively, that were humans surely bigger than skinny little Cricket. The big figures, objectively, that were the dogs, the Bioforms. He'd heard that, back in the day, there were more kinds. Perhaps, elsewhere in the world, there still were. Cats and rats and lizards and who knew what else? The Factory just made dogs, though. Probably that was for the best. Cricket didn't want to spend his days digging up rat nests and collecting eggs.

One of the four big chimneys was smoking, a lazy trail of black smudging up into the low hanging cloud. It could have been anything, of course. Wood for warmth. Old rubbish. Cricket knew in his heart it wasn't, though. When the black smoke rose from the Factory chimneys it was the failures they were disposing of. That was what everyone said.

The doors were solid, metal shod, of a newer vintage than the Factory itself. Back in the day, an alliance of Bunker factions had tried to take the Factory over and own its dogs. The putative claimants had regretted their hubris sure enough, but they'd used some portable artillery to stave the door in before the furious resistance of armed dogs had hounded them off across the hills. The new door had been a gift from one of the villages, brought by monks like Cricket a generation ago. The Factory got a lot of gifts, just like the surviving Bunkers did. Not quite taxes, not quite extortion. Just an acknowledgement by those who scrabbled at the land, of those who held the guns or had the dogs.

And us? Cricket wondered. If the world was a machine of awkward, ill-fitting parts, then the Apiary was trying to be the grease between them, so that everything could just keep moving one more day at a time.

You could get crushed, between the pieces of a machine. Looking up at the Factory's bleak walls, Cricket was all too aware of it.

The doors swung open. The old ones, the destroyed ones, had moved on their own. There were two dogs behind the new ones, pushing them with brute force. But then brute force was something that the dogs had in abundance. A perfect knife-edge combination of the natural and the artificial, that would always be more efficient than just building a machine thing from scratch. The Old had possessed many machines to do all kinds of jobs, Cricket knew, but few of them remained in operation now, certainly outside the Factory and the best-preserved Bunkers. The dogs could turn their huge hands to all manner of tasks, though, in a way the old machines had never been able to do. They could make their own decisions, work within a human context, heal their injuries. The Prior said they were the pinnacle of the Old, miraculously maintained in these lesser days of the New, or at least he did when there was any chance his words might get back to the Factory. Looking at the dogs, Cricket felt it was selling them short.

A man, stocky, broad-shouldered, a lot bigger than Cricket, came out to greet them. He had a mild brown eye and a cold glass one, like a smaller version of the red lens set into the dog's skull. Through both glinting orbs, the Factory would be watching. The management that controlled the place and made decisions for its running. Every one of their followers was a proxy for their presence. It was another reason the Factory had thrown back the various forays of the Bunker forces back in the day.

"Good afternoon," Cricket said, aware that the day was

shading into evening, really. The business with the wheel was going to see him walking the track in pitch darkness, or worse.

"Bellman," said the Factory man. "Willem Bellman, ain't it." He had an accent, a man who'd grown up in one of the villages of the West Edge, where they spoke in more of a drawn-out way, like their vowels were digging their heels in. "And you're Siblen...?"

"Cricket." A small bob, keeping his hood up and the tall collar of his under-robe zipped, so that nothing of his face could be seen save the eyes. The mystique of the Apiary, as he'd been taught. Enough to give a village bravo pause, perhaps, before laying hands on a monk or pushing one around. Probably not anything the Factory people were going to be spooked by. But Bellman, despite his size, didn't look the pushy type. He smiled easily and with all his face save that glass eye. Then he stepped forward and reached up to scratch under the chin of the dog who'd come with Cricket. Just right into the huge creature's shadow, fingers within inches of those fangs. Not a second's hesitation. There had been a dog, once, with one of the Bunkers. Maybe one stolen from the Factory, or a renegade who'd run off. Maybe one come from elsewhere, some other place they made dogs. It had been mad, though. They'd kept it on a chain, let it slip only when there was an enemy in sight. They'd had to kill it in the end, Cricket had heard, because it was more of a danger to them than a deterrent to their foes. He'd seen it, the once. Seen the people who'd had it on a leash, half-strangled; the whips and the spiked clubs to keep it in line. Seen how they were terrified of it. A generation before, one of the villages had a dog, they said. Worked hard, carried loads no human could and knew with a sniff whether any stranger was there for good or ill, but it had lived outside the village proper, and everyone had kept their distance for all the years of its life. Scared, because if it had a mind to it could have killed every human being in that village. And here was Willem Bellman just scratching this dog behind the ears like it was any village pet or working animal.

There had been a moment, on the road. When the dog had stopped, and looked at him a new way. It had taken all Cricket's self-possession not to soil his robes, honestly. The dog, suddenly recast as predator, through nothing more than a change in the way it held itself. He could never have stood easily, so close to it. Even though he was still well within its reach now, given those long arms and clawed hands.

Beyond the Factory doors there was an open space. An exercise yard, a muster point, some such. Right now, it was just a place for humans and dogs to cross, carrying or pulling loads. There were workshops where ruddy forge-light mingled with the harsh white of welding, and hammers lifted by hand sounded their hard melody against the hiss and stamp of machines. There was the main factory building, the doors of which did not open to admit him. Nobody got to see the masters of the Factory, ever. Nobody needed to, given they could speak to their servants at a distance, and see through their implanted eyes. Not as though Cricket was going to launch a one-man assault on the place, but he could have been a saboteur. He could have been a spy. Since the fall of the Old, the Factory had been coveted and resented by many of the other surviving powers of the world. They weren't the trusting sort.

"Tribute, from Brokebridge, Mister Bellman." Judging that the man wouldn't be offended by the standard all-purpose term of respect, the way some of the Bunker-folk were.

"That's good, that's good," Bellman said. He looked

over the pups in the wagon, visibly counting, yet also just providing a point of view for the true count that would go into the Factory records. *Brokebridge, eleven pups*, recorded in the ledgers. Enough, perhaps, to see some aid to the village from the Factory, in a good year. Enough, perhaps, to see the village spared the Factory's collectors, in a lean one. The brittle economy of the New.

"I'll just be..." Cricket said, looking back through the gates at the outside world. It was dark, that outside world. No electric lights, and a moil of cloud above, as the Prior would say. Cloud upon cloud pulling double shifts to darken everybody's nights. Beyond the wan light shed from the Factory's slot windows, he'd have to keep to the track by the feel of it through his boot soles.

"You'll have some soup, Siblen," Bellman said.

Cricket started. Would he? He wasn't sure if it was a question. Wasn't sure if it wasn't some way of getting him to walk into a bad place, a trick. As though the chief ingredient in the soup might be one incautious monk. And if they'd wanted monk soup, then the big dog who'd come in with him could have carried him to the pot in one hand, but somehow there still seemed a virtue in getting him to walk to his own doom. As though they were things from one of the bad stories for children, and the soup would taste the sweeter for them having fooled him. But all that was in Cricket's head. Nothing in Willem Bellman's broad, honest face, nothing in the brooding quiet of the dog, nothing even in the eerie glint of their glass lenses. Just Cricket letting his superstitious side get the better of him, as though he was no more than a credulous village swain.

"Soup," he said, "would be very welcome, thank you, Mister Bellman." Casting another glance at the great gates, which were now being hauled closed by the two attendant dogs. All the Factory dogs were different, he saw. Breeds, sizes, builds. As though they'd had to work at bringing out the different human-ish characteristics that had been within each pup. Lean dogs, stooped high-hackled dogs that were more comfortable on all fours, big burly masses of muscle like the dog he'd arrived with.

"We'll find you a mattress, Siblen," Bellman said, reaching to tug at Cricket's hem and then drawing back, gesturing instead. Apparently a little of the monastic mystique did travel as far as the gates of the Factory. "You won't want to be travelling tonight."

"Very kind." And there was a way you were taught to be, outside the monastery. Cool, distant, detached from things. Otherworldly. Which was a weird agony for Cricket, whose instinct was always to fill any silence with blather, and who rather desperately wanted to feel that people liked him. But to be a monk was to be respected, ideally, not liked. And so he was supposed to be haughty, and only nod a tiny degree, and snoot at people down his nose, save that of all the people he could see right now, there wasn't one he could have looked down on. Not even the dogs on all fours.

Bellman led him into the main factory building, into a big refectory hall where men and women and dogs were all sitting on benches, being served soup. The dogs and the humans got the same soup, but the dogs got more of it. Plus, the smallest Bioform dog Cricket ever saw went down the line and each dog got a little selection of chalky-looking pills too, as did a few of the humans. Medicines, he recognised. Things to counteract any pushback from nature, over what had been done to them.

He sat down, Bellman to one side, the big dog – Deacon, he had learned – to the other, feeling like he'd been put down a well with barely a glimmer of light from above. Nobody had to tell the servers to bring him a bowl, of course. They just knew, because Bellman knew, which meant the Factory knew.

The soup was very good. The Apiary mostly resorted to soups and stews, save on feast days, because it was a good way to feed a lot of mouths with no waste. Those soups were thinner than the Factory workers got, though. Cricket couldn't quite get to the bottom of the bowl, despite all the appetite the day had worked up in him. When he finally flagged, Deacon was more than happy to take the last lees of it from him and lick the bowl clean.

After that, the promised mattress was found – not in the big dormitories where the Factory people slept, humans and dogs together, but in a small square room they perhaps kept for such guests, that was still bigger than the cell Cricket had back at the Apiary. Bellman paused in the doorway, and Cricket sensed a moment's almost prurient curiosity in the man. The monk had unbuttoned his collar to eat, of course, folding it down over his shoulders the way he'd have worn it within the cloisters of the monastery. Even with that, and his hood pulled back, Cricket's close-shorn head and youthful face offered little in the way of clues. Another Apiary tradition, that its messengers be sexless as the bees themselves. Cut off from the human, as befitted the power they were emissaries of. And Bellman had kept up a companionable talk at the dinner table, asking about villages by name, and travel conditions, and very little about the Apiary itself, as was proper. Enough that Cricket's own predilection for over-talking had been kept in check, and the dignity of his role preserved. Which had, plainly, left plenty of questions unanswered in the man's mind. *Is it true that...?* And who knew what stories they told of the monks, here in the Factory. Just as many as they told about the Factory in the Bunkers, or about the Bunkers in the Villages. The fall of the world had broken everything into pieces. *We are all born strangers into the world*, as the Prior said.

Cricket waited for some intrusive question, something that pried too far into the traditions of the Apiary, that he'd have to be firm about rebuffing. Or worse, something *personal*. For, even though Cricket sometimes longed just to be able to *talk* to someone from outside the cloisters, it wasn't done. It gave people ideas. It eroded the sanctity of the Apiary, risked all they had built. And Cricket felt, simultaneously, that he might quite have liked being friends with someone like Willem Bellman. And at the same time felt that Bellman, big, confident and with the glass eye of the Factory set into his head, was somehow a terrifying figure, and might suddenly lay hands on him. Complicated things.

But Bellman obviously found his thoughts too taxing to set to words, for he just shrugged. "You'll be woke by all the others first light, I've no doubt," he said, almost apologetically, as though in the normal way of things monks rose languorously at noon. "But I'll look in on you if not." And backed out, as though Cricket really was some frightening supernatural entity, capable of eldritchness and witchery.

The next day, after a solid night's sleep built mostly on exhaustion, they had the wagon ready and empty, cleaned so that barely a whiff of over-excited pup remained to it and the cracked axle bound up with rawhide by some factory tinker. A little package of wafer-like bread, too, or wafer stuff that was more like bread than anything else. A jug of water for Cricket to refill his plastic bottle from. A warning that they'd seen scavengers from one of the Bunkers in eyeshot of the walls overnight – the eyes of the Factory being proof against the dark when they needed to be.

"I'll be careful," Cricket promised, although honestly the country was light on cover all the way back to Brokebridge, and if some Bunker warriors caught him he'd have to hope the old monastery glamour and his lack of anything pillage-able was enough to keep him safe. Still he kept to the track and kept his hood and collar up. A weird little silhouette against the flat, white morning sky, some gnome of folklore rather than just a little man in several layers of robes. And the Bunker soldiers didn't stay him, or even appear. Instead, on the road, close to where the wagon had thrown its wheel the previous day, there were birds. Black, ragged swatches of feathers circling and dropping and squabbling amongst themselves over something close by the track in the long grass.

He should never have looked, but curiosity had an odd place in monastery teachings. They were unworldly, which meant things like this – earthy, worldly things – were supposed to be beneath their notice. At the same time, they were custodians of knowledge, meaning they were supposed to want to *know*.

Cricket wanted to know. He'd always fallen more towards that pole of the spectrum. Something was dead out there, and if it was a person, someone should report it. He might even be capable of bringing the corpse to Brokebridge on the wagon, if he was feeling sufficiently full of soup and wafer-bread to put in the effort. And if it was an animal, well, possibly that should be reported, or maybe if it was fresh he could even load it up and bring it in for the pot. Waste not, want not.

It wasn't an animal, neither natural nor Bioform. It wasn't a human either, or not any human Cricket had ever seen before. It was a prodigy, a monster. Truly a *monster*, in that sense the Prior taught. *Monstrum*, a warning of terrible things to come, by its very appearance. A thing from elsewhere and otherwise and old stories, lying there dead as though it had fallen from the sky.

5

IRAE

Irae, from the Latin. *Wrath*. You all must have your little joke.

You don't know how complicated it is, Wells. Being here, where we shouldn't. Yes yes, I volunteered. Too late now for second thoughts, you say. Months too late and a hundred million kilometres too far. Still, with me, second thoughts can come late. Some of us take a while to digest things.

The locals here, they only have one mode, Wells. They just *are*. Can you even imagine? The simplicity of being just yourself, and never having to shift gear. No of course you can't imagine, doggo. You only have two modes yourself. You're a binary being, Wells. Standard operations and economy mode, and the latter is only really suited for the sort of conditions they don't even *have* here. The sort of conditions we don't even have back home any more unless you get really far out or high up. Honestly, they only build us that way for the sort of emergencies we've not had in a generation. If any of us really needed to go that far into the dust, back home, we'd just wear a suit, wouldn't we? No need for us all to be so damn *rugged*.

Speaking of rugged, Wells, I know they did a lot of work on us, before we came out here: bones, blood, respiratory system, but that doesn't stop you feeling the drag, does it?

Like weights on every limb. Unless I dial myself up until I'm fizzing so much I don't notice the weight.

Wells' channel: *Have you found Tecumo?*

My channel: *I'm looking at him.*

And I've never been able to put emotional subtext into my speech properly, and the channel link washes it out anyway. So you can't hear, Wells, but I'm being grim. It's black humour, Wells. And I wait for you to pick up on it and appreciate my inner complexity, but circumstances are adverse, and none of you look for it, in me, and so you have to be all pedestrian about it and just ask:

Wells' channel: *And?*

'And' indeed. For there is Tecumo. My tongue, tasting the air, picks up his savour, which is mostly *clean* and still slightly *home* and *ship* and nothing at all like anything from here. But fainter than it should be. Unsurprisingly.

And the very first suggestion of decay. We've been looking for him all night after all; he's not fresh dead.

I let myself move the slider of my metabolism, shift gear towards life in the fast lane, rather than life in cold blood. The coils in my vest circle with a current that fights to pass through the thin metal. That fight makes warmth. The warmth touches my blood, my brain. The world around me, that was clipping along at a dangerously brisk pace, slows considerably down. My picture of the world sharpens, and I can properly itemise and appreciate exactly what's before me.

My channel: *Dead since yesterday. Plus I spy one murder suspect still camping the body.*

Wells' channel: *Armed? Are you in danger? I'm on my way.*

My channel: *Still your heart, doggo. They don't even know I'm here.*

I dial back on the current, letting the world run ahead of me. All very well to be living fast and thinking hard, pushing the very edge of goblin mode, but if there's anyone out there with thermographics then I'll start shining like a little star. And you'd tell me, Wells, that nobody around here's using that kind of kit, and Tecumo would have said just the same, and now he's dead.

Compared to Tecumo's sad corpse, Murder Suspect Number One blazes with life to my pit receptors. A nice juicy little mammal with a high-end metabolism. If it came to a foot race I'd have to dial myself up significantly to catch them. But they're too busy looking at what they've done to notice me, and so I could save a lot of energy by just slithering up close and playing the ambush predator, favourite role of my un-Bioformed ancestors. It's tempting.

My channel: *Looks like... I think a local got him.*

Wells' channel: *Why do you think that, Irae?* Now Wells, *you* can put a lot of character over the airwaves, but that's because the bottle is always full, with you. Honestly, face to face you're exhausting. You need to learn a little chill.

My channel: *I think that, Wells, because Tecumo's been got and because there's a local standing gormlessly over the body. I think it was a damn fool thing for Tecumo to go it alone and we should have put the fear of all their primitive belief systems into these local savages and turned up guns blazing.* You see, a reasoned proposal, a well-thought-through plan. Chill.

Wells' channel: *Correlation and causation. Do not engage the local.*

And technically Wells is in charge when it comes to fighting or not fighting the locals, but if I get close and kind of reveal myself in just the right way, maybe the local will pull a weapon on me and I can heroically murder them in self-defence. One can always hope.

Irae. *Wrath.* But I don't get angry, Wells, you know that. Not even when our leader's just been offed by some ape-man.

I've been moving in on our killer all this time, Wells. Partly because you said *do not engage* but you didn't say keep your distance. The grass is long here, the fields left fallow. A thousand small things move all around me, but when I'm dialled down and slow it means I'm focused, too. On my prey.

Not my prey. Only the Person of Interest; that good enough for you, Wells?

A small local human suspect, but only by local standards. Entirely average for home. Tecumo was no bigger.

Wells' channel: *We're incoming.*

I'm very still. Suspect is moving. They have a bad-tasting little cart thing, and I think they're going to dispose of the body at first. But they leave dead Tecumo lying there and just return to their cart to make a getaway. To my eyes, dialled down as I am, they fair hightail it off through the grass to the track. I have enough thought left to understand that honestly it's more of a trudge. Definitely wanting to put distance between themselves and their misdeeds.

Yes, Wells, their *maybe* misdeeds. Or maybe they just doesn't like the company of corpses.

Wells' channel: *Hold for us.*

My channel: *Body location here.* Dispatching the coordinates direct from my locator so I don't need to think through the numbers. *I'm following the killer.*

You don't like that, Wells. If I had more mindspace, I could imagine you and Ada arguing over what to do, during which time I slink after the cart and the maybe-murderer, keeping to the grass. I know Ada's won you over when I hear:

Wells' channel: *All right then but don't—*

My channel: *Engage, I know.* Putting a definite stop to the conversation before you can start adding any extra qualifications.

The maybe-killer hauls on the cart and moves off. I creep through the grass, stalking. I wonder how close I can get to their heels before they know I'm there.

6

CRICKET

The corpse.

Cricket stared down at it, his stomach shifting queasily. He'd seen dead bodies before, of course. Old monks dying in bed. Monks of all ages dying of sickness. The violent breach of accident – a fall, a heavy load inadequately secured, that sort of thing. And violence. This was a body dead of violence, for certain. He knew a gunshot wound when he saw it, because he'd been a little kid when the Bunker Wars were on, and back in those days you saw bodies whenever you went outside to scavenge for food. And it wasn't like everyone considerately handed back their guns after the Bunkers settled their differences. A settlement that had resulted in only one big force remaining in the area, a single Bunker with its complement of warriors to dominate everywhere the Factory's shadow didn't fall. The others had been filled in, rooted out, routed, and those who'd come in from elsewhere had been driven back home, or left after signing treaties. Meaning another round of violence in the future was entirely probable.

So, on encountering a random body displaying clear signs of gunplay, Cricket would normally have an escalating variety of guesses. A feud settled the old-fashioned way, perhaps. Or maybe brigands, because there were still loose soldiers wandering about like those old stories, deprived of their masters and preying on the weak. And yes, the stories went long on how noble those masterless men were, but in Cricket's experience a Ronin was just a bandit who'd once killed people to order. But such threats were a fact of life, and the Apiary's mystique worked hard to breed sufficient superstition that lone monks could generally travel safely. It wasn't as though they had much worth taking, after all.

Worse than brigandage would be another outbreak of inter-Bunker war. Some new armed force intruding from outside, ready to contest the region with the incumbents if it found them wanting. Because although those conflicts were one group of armed people conflicting with another, the worst always seemed to be suffered by those with fewest guns and the most interest in peace.

But this wasn't that. Was Cricket relieved? He was not.

The body was chalk pale. Or, no, not even that. One of the monks at the Apiary was an albino, after all. It wasn't anything sinister. The skin tone of this corpse was weirdly ashy, as though there was a dark, almost metallic layer right below the dead pale skin. If it had been just that, Cricket might have put it down to cause of death or some post-mortem oddity. The face, though...

The eyes were very small. The nose was very large, nostrils bristling with hair. And Cricket wouldn't say he noticed people's nose hair, particularly, but this was virtually an internal moustache's worth of it. The mouth was narrow, lipless. The face as a whole was weirdly flat save for that rudder of a nose. Not flat like some human faces were. Eerily level from forehead to chin, and the eyes just set in little dents. And something weird about the eyes. Staring open, but one of them winking at him. Unwillingly he bent closer, squinting until he made out the translucent eyelid stretched halfway down over the dead orb. And the

Apiary preserved enough records and recordings of the Old to know that people could look all sorts of different ways, but he didn't reckon any of them had two sets of eyelids.

He straightened up and stepped back, spine crawling. Staring down at the dead... man? Woman? Monster? Probably man but he couldn't be certain. The face was far enough from any familiar markers so as to be sexless. The hair – silvery grey, almost metallic – was rough-cut, not short, not long. The body itself was small – standing, the dead thing would be no bigger than Cricket's own scrawny frame. And its clothes were as weird as its face.

He didn't even know what they were made of, he realised. Not wool or linen. Something more like the uniforms and recovered antiques of the Old, made from fabrics that didn't really rot. Plasticky, he reckoned. But some of it was metal. Some of it was machine-y. There was gleaming silver set into the cadaver's skin, poking out of collar and cuffs. Discs of it, and ridges beneath the pallor that might be a whole system of rods and things. He forced himself to kneel down. He didn't recognise any of it, but this creature was plainly from somewhere like the Dog Factory, where they still had working remnants of Old tech.

The world around him was very quiet, as though paying its respects.

Or, looked at from a less poetic standpoint, as though something dangerous was out there, and all the little things of the grass didn't want to draw attention to themselves. Lucky for them that Cricket was here to be large and conspicuous in comparison.

He stole a look around. There was nothing. But the grass was tall and *someone* shot this monster.

He forced himself to touch the corpse. There was no residual warmth to it. So: not dead recently. Dead overnight.

Dead since yesterday maybe. A chill ran down his spine, that this corpse could even have been lying there as he'd passed this way with the pups.

Those small, dead, utterly inhuman eyes bored into him. Less kin to his own than the Factory dogs', or a sheep's. A thing from Somewhere Else far more than any raiders from a distant Bunker or some coalition of foreign towns.

Cricket sat back, wanting very much to just run away but knowing he had a duty to the Apiary. This was knowledge. This was something new. Prior Stick needed to know. Really, he should take the entire body back to the monastery, but that wasn't going to happen – squeamishness and a lack of raw muscle power informed him of that. Instead, he fumbled about inside his scrip until he found his little device. It was a wooden box, scratched and battered and shabby, nothing any brigand would bother with, or that was the idea. Inside, once he'd fiddled the secret catches about, was a piece of the Old. He lifted it to his eye and pointed it at the body, took a dozen views of it, close on the face, further, to take the whole thing in. Then he robbed it. And it was a dead monster; probably there were no dead monster authorities who'd come after him for the theft. Still, he felt guilty, because that was the way he'd been brought up. Felt guilty, but still took the three machines that were clipped to the monster's plasticky belt. Marvelling at the mechanism of the clips, which was elegant and beautifully crafted, yet simple enough for him to work it out at once. In the end he took the whole belt, because those clips were a work of art and maybe the Apiary could duplicate them. Cricket was all about the small advances. He found they tended to get fewer people killed.

Then he stood and rather belatedly put a good few paces between himself and the corpse. Nobody was around to see him, or at least he saw no-one. He was well aware of how it looked, if someone caught you with a dead body. The monastery had preserved those stories, too. People always jumped to the worst conclusion in such circumstances, no matter how much evidence there was to exonerate the person found bending over the corpse. Especially if that person had the corpse's possessions. That also looked very bad, insofar as Cricket was any judge.

By then, the unnaturally protracted quiet was really getting to him. He'd been trying to tell himself that it was just his own blundering about that was spooking all the little things, but he'd been quiet still at the corpse for a while, and not a bird was piping up. It was a particularly fraught quiet that mere human beings didn't tend to generate, but he'd heard it around the Bioform dogs. A taut silence when nature is tacitly acknowledging the top predator.

Cricket, a long way from anyone's top predator, practically picked up his skirts to scurry back to the wagon with his ill-gotten gains. Not a thief, he told himself. Not really. A recoverer of artifacts. Much more respectable.

He hauled on the wagon – easier now its bed was empty. The Apiary was a couple of days walk from here, sitting on the far side of Brokebridge by Clearwater, at the notional fulcrum point of the region where they could stick their noses into everyone's business. The Factory, the Bunker, all the major farms and settlements. Two days of brisk walking, a night of sleeping rough or begging for shelter at some farm or other. Enough to put all that unpleasantness well behind him.

Except, when he walked off, he had a definite sense of the unpleasantness keeping pace. The quiet drifted along at his heels like an invisible hound. The birdcalls he heard from up ahead went still before he passed. The rabbits that had been warily cropping grass were gone before he neared them. Even the insects, the chirr of his namesake, they went still. All the world looked past his heels aghast and held its breath. Until he couldn't really pretend it was *him* that was making poor nature tremble so.

Something in Cricket's head said *run*, but where, exactly? And there was nothing. He looked around, obviously. He peered about covertly. He pricked up his ears and listened into that uncanny quiet. Nothing. Not the least motion through the grass. Not a scuff of a footstep, the click of a firearm. Except every time he stopped to scan for his pursuer, the absolute nothing that he saw and heard was, definitively, just that little bit closer.

If he wasn't going to run, then that same lunatic part of him told him he should turn, shout, rage, brandish a stick. Scare off the thing that he couldn't know was there, that he most definitely knew *was* there. Send it running invisibly for the hills on soundless feet. Except how would he even know if that had worked, exactly?

He turned his face forwards, put the wagon rattling at his heels, and walked. From that point onwards he didn't stop, didn't look round. The thing – the utterly undetectable thing – would follow. Would creep up on his heels like children playing that footsteps game. And, like someone who didn't wish to play, he wouldn't turn round, wouldn't stop. Would just walk on and hum to himself to blot out the deafening roar of the silence.

Incredibly – exhaustingly – he kept this up until dusk was threatening. By that time he felt that he'd been running a race all day, because the tension of the world had never let up, and nor had the clench inside him. Somehow just being calm and measured and not giving in to the fear had worn him down more than running away screaming. The one-sided détente he had established with the invisible monster and killer-of-monsters was going to kill him through sheer nerves.

And, seeing that darkening of the eastern sky, he realised he couldn't keep it up. His options had dwindled to almost none. He could not just keep walking in the dark as though there wasn't a monster creeping soundless at his heels, for surely the dark was the thing the monster that wasn't there had been waiting for. Nor could he just pull the wagon to the side of the road and tilt it on its side, to huddle in its shadow. How could he sleep, with that monster not out there, not about to pounce on him the moment he closed his eyes? No, no, quite impossible.

He came to the vice of this decision just as he reached a fork in the path. There had been something here, in the Old. The foundations of it were still visible, and under other circumstances he'd have made it his camp, where the cracked concrete could keep him up out of the worst of the damp. There were several trails that converged here, for no reason other than it was a landmark that all their makers had originally aimed for. One went straight to the Apiary, but another led to Brokebridge, and Brokebridge was close enough he could reach it before full dark.

Just a town. A village, really. Several hundred souls, reclaimed farmland where it was worth tilling and goats to go where it wasn't. Even an old generator that they could shovel their waste into, to get some electricity going on special occasions. But right now Cricket would have taken a shack full of wandering brigands over whatever the hell was out there on his trail.

What he got, not long after diverting his course towards Brokebridge, was a herdsman. Herdsperson anyway, hard to tell in the gathering gloom and the all-encompassing plastic poncho they were wearing. The wavering beam of a lamp suggested their farming clade was quite a well-off one: electric, with some of the old rechargeable long-service batteries involved. Quite a piece of kit to take off looking for stray animals. Which intent was given by the crooked stick they had in their other hand, and the dog at their heels. Not a Bioform dog, of course, which would have had to crunch down exceedingly small to count as being at anyone's heels. A big, calm working dog, though; the sort the farms and villages bred and got good use of, and tithed the spare pups to the Factory.

The herder shone their torch into Cricket's face suspiciously, seeing no more than a tall-hooded shape on the track. A moment's uncertainty and they registered his clothing, and bobbed respectfully.

"Siblen," they said. "Out late. Not seen a goat, have you?"

"I have not," Cricket said apologetically. He reckoned he could even guess the herder's family, given a couple of tries. He knew this country and its people, after all. The slow-burning panic of before was starting to gutter out. "Tell me, how might—?"

The dog growled, its hackles up, its lips drawn back from its teeth. At first he thought it was growling at him. Then he understood, and the skin between his shoulder blades began itching. The herder bent down to remonstrate with the animal, then paused, looking out, flashing the torch around.

"Don't know what's got into her," they said uncertainly. The dog was still keeping up that low ragged note, but it was backing off, too, tail between its legs. And the working dogs around here were renowned for their tenacious courage when their herds or masters were threatened.

"You come alone, Siblen...?" the herder said nervously,

and then froze, torch held out, the beam lancing into the dark. They gave a high, shrill sound, practically falsetto. In that moment Cricket understood that they were male, and, simultaneously, had been utterly unmanned by what they had seen.

They fled. Not a word, just turning in a swirl of poncho to dash off into the dark, the dog whimpering at their heels. Cricket called out after them, then tried to pursue them, but they were lurching off the track and away, and the dark was gathering. Soon after, he found himself just in the middle of some scrubby pasture, the track lost, the herder lost. The only company he was still in possession of was whatever had been dogging his footsteps since the unnatural corpse.

He turned all the way round, and then kept turning, looking at the dark. He wished he had a torch. There was a little flint lighter in his scrip, that was bright translucent plastic, but he didn't think half a candle's worth of flame was going to do anything but pinpoint his own location for the stalker. Not, he suspected, that it needed the help.

"I am a monk of the Apiary!" he declared. "I walk in the protection of our master! You dare not harm me, for fear of the retribution of Bees!"

And, from the dark, a deep, croaking chuckle. It made his blood run cold. Because whatever made that sound was plainly nothing like a human throat, and yet was working hard to make a human sound. And when the Factory dogs talked, you could hear that their throats had never been meant to form words before the procedures they'd gone through, but compared to this they were Cricket's close cousins.

His own voice dried up in his throat and he fled. Just away from that sound, that was the important thing. Any other benefit of his destination was just a bonus. And, because the universe gave as well as taking away, he saw lights almost immediately. High lights, quite far, but he could see the broad shadow they were set atop. It was the wall of Brokebridge, the town. He was exactly where he needed to be. And they'd have closed the gates, but they always kept a watch. He was a monk. They'd let him in.

His nerve broke when he got a little closer to the walls. Or broke further, given that he was running flat out by then. Shouting, waving his hands, all monastic dignity thrown to the winds. "Let me in! I'm a monk! Open the gates, please!" And, joy of joys, movement along the top of the wall, that grand heaping construction that was jagged plastic and stone and the rubble of the Old, cladding a stout palisade of girders and posts. A ward against beasts and brigands both, because the nights of the New were not safe, as Cricket would absolutely attest to right then.

He heard voices, and then someone – bless them! – lit the big electric lamp over the gate, burning up precious stored battery so that he had somewhere to head for. That pool of pale gold, illuminating the track, the wall, the great metal-shod gates even then creaking open. There were people beyond, with guns just in case this was some raider's trick, but the terror in his voice had obviously been persuasive.

Once he was inside, and safe, and no sign of any pursuing monster, he'd have a lot of explaining to do. He'd have to work hard to stitch together the rents he'd just made in the Apiary's imperturbability. But Cricket reckoned he was up to the task. He'd take it as penance, in return for being saved.

The thing. The monster. It was right at his heels he was sure. He didn't look back. He felt that even seeing its shadow, its outline, would strike him down. His legs would fail and he'd be lost to the dark. He just put on an extra burst of speed towards the gates.

Someone up on the wall hailed him, then there was an excited babble and he distinctly caught the words, "What's that behind them?" and "Run, Siblen, run!" as though he wasn't already.

He was going to make it. He was just staggering into that welcome splash of light, running up the shallows of it towards the brightest glare just before the open gates. The thing behind him was shrinking from the illumination, surely it was. A creature that existed only in the fear and the shadows. He, Cricket, was going to live!

He was close enough to see the wide eyes of the townsfolk at the gate, the shake of their rifle barrels, the individual hairs on the beards of the men, when the *other* monster lunged from along the line of the wall, extended a horrifically long, hairy arm, and snatched him away into the darkness.

7

WELLS

Irae's channel: *It's like he never saw an alien before.*

My channel: *We're not aliens.*

Irae's channel: *We're not from round here. Maybe we should have worn big fishbowl helmets with antennae. Like in that recording. The one with the gorilla-form.*

My channel: *That was a human dressed as a gorilla.*

Irae's channel: *Not when they remade it.*

Ada's channel: *Enough from the pair of you. I think he's coming around now. What did you do to him, Wells?*

My channel: *Me? It was—*

Ada's channel: *Who grabbed him, precisely?*

I'm getting het up and stressed again, with all this interrogation. And Ada does not understand how *hard* this is, for me. Neither of them do. She is, after all, only human, and Irae can just dial down until it's so calm you could poke it with a stick. For me, though, it's hard. We trained so intensively for this, took so many precautions, medicines, supplements, surgery, training in the bright room, in the swinging pendulum. They never told us about how it all impacts on the *senses*.

My channel: *He fainted. That's all. I didn't do anything. I went and got him like you wanted. I brought him to you. One thief.*

Irae's channel: *And murderer.*

My channel/Ada's channel: (Simultaneous.) *We don't know that.*

Irae's channel: (Snickering. Or at least an electronic sound meant to that effect.)

The small thief is indeed regaining consciousness. At Ada's gesture, Irae and I absent ourselves. I'm not sure she's thought this through, really. How it will look and what the native will think. What we might get out of this. But I do what she says. And it's not like *that*. It's not Humaniforms giving the orders and we 'lower Bioforms' saying yes. It was always the three of us, Tecumo, Ada and me. Irae not so much, but the three of us with the dream of seeing the old homeworld. And then the message came, and suddenly there was our opportunity. People had been talking about it for a generation, at least. Once we'd stabilised ourselves after the Crash, pulled together, made sure we were sustainable. Still amazing, that: in the most inhospitable but inhabited place in the solar system, we had it easier than most of Earth.

And besides, not as inhospitable as it was. Still plenty of ways to bite you, my old home, but we've dulled her teeth over time.

I hear a groan from the little native. Irae is… somewhere. I can tell the general area it has gone to ground in, but it's good enough at hiding that I'd have to physically go over and start nosing about before I found it.

'*It*' is, I know, generally considered insulting, as a way to refer to a sapient being. But it is also how Irae signals that it honestly cannot be bothered with everyone else's interest in the subject.

Our encampment is plainly somewhere that locals occasionally use too. A dip in the land where three slabs of crumbling concrete have been leant together – long enough ago that the joins between them are seamed with moss, a whole miniature ecosystem that grows and dies and bustles and whispers and—

I break from the thoughts, irritated at myself. Every little moment, every sound, every scent, is like a leash tugging my brain this way and that. I have to focus. I don't think the native is the one who killed Tecumo, just some opportunistic looter, but who knows? What happens if I'm tranced out to all the everything that's going on around us and Ada gets killed too?

I was supposed to be watching over Tecumo, after all. It was my watch he died on. Ada hasn't accused me, pointed the finger, *Bad Dog*, but she doesn't have to.

Ada has some of the local food we unearthed. To offer to our prisoner as a bribe for good behaviour. Not to eat herself. It was in tins but we don't trust it without analysis. We found it shallowly buried under the makeshift shelter of the slabs. Under the shadow of where they lean together. Presumably this is some mutual aid ritual. You're in need, you come and take. You have spare, you come and bury. A sensible system, and it tells me the locals have at least some wider sense of community. Some of what we heard in the Signal suggested things were a lot worse.

Tecumo is dead. Things are a lot worse. Dead because he was too confident. Dead because he wanted not to spook the locals by bringing me inside their walls. Dead because I wouldn't insist. Wouldn't just go with him. And then, listening on his channel, there would have been clues, signs that something was about to happen. And I didn't catch them because of all of it, the hugeness and the richness and the constant insistent intrusion of it. I did badly. I am not like the first Bioforms, the servant-soldier-slave ones, where it was all *Good Dog, Bad Dog* to yank me around and get me to do what my masters want. I have no masters, only comrades in this, our venture, that suddenly looks so grim and doomed. But I am more than capable of knowing myself a *Bad Dog*.

I want to just go be right beside Ada. Not hide here, in trees, crouched low. Only a two second run from her side, even in the punishing gravity of Earth, but two seconds is enough time to die in. Back home, two seconds is more than enough. Even half-tamed, that world can kill you plenty of ways if your equipment fails. Earth is supposed to be safer. The world that, however modified we were, is still the one best suited to us. Except it just means all the wariness we have, for the ways home can kill, is useless here. Earth kills you in new ways. Like guns.

I narrow my eyes, cock my ears, focus ferociously on the native as he sits up. We took off him anything that seemed like a weapon, but the first sign of aggression and I will be there. I will tear his throat out, in revenge for Tecumo.

I seem to hear Irae's snicker again, but there's no activity on its channel. Just the little sarcastic Irae in my head.

The human native is around the same size as the regular people I'm used to back home. After the Crash they tweaked them a little, because we were short of resources. Tweaked them, and got selective about what Bioform models to engineer. A lot of Irae's model, back in those days, because of the way they could dial their energy consumption up and down. Some of those are still around, too, the most veteran of old-timers, because a few of the dragon-model designs turned out to have real long lifespans if they kept themselves dialled low and slow.

The thief sits up. His hood, which was weirdly pointy, stiffened with wire, has fallen back. I see a strangely angular face with big eyes that make him seem helpless and vulnerable. An old, old instinct kicks in me, that here is something I should be defending, not watching as a threat. It's a reaction that humans and most of my kindred models have, to small things with big eyes. Pups, babies, cartoon characters. The mental wiring that had humans brought up by wolves and wolves domesticated by humans, someone once told me. But here it's misplaced. Wolves are threats, after all, not like dogs. Those big eyes are just the natural human size, and I'm used to the smaller orbs of our local Humaniforms, that are modified for low pressure and dust exclusion. This is no baby-faced innocent. A thief, maybe a killer, certainly a dangerous inhabitant of a wild and lawless planet.

Lawless. Maybe that's the problem with Earth after the Crash. That you *can* live there easily. Back home we had to hold together. If anything stopped working we'd be out of food, and maybe atmosphere too. And we had help, to make sure we got through it. Help coming out of the wilderness – not that anyone's a *native* back home, but help from the one thing that could just about claim it. The oldest resident of the red planet, the first colonist of the dead sands.

I'm drifting again, but I'm brought back sharply when the native rubs his shaven head and then sees Ada.

Now the point about me and Irae getting clear was not to scare the local. That turns out to have been a waste of time because, when he sees her, he just yelps and scrabbles back against the slope of the concrete slab. Hands thrust out towards her so that I almost break cover and rush at him. Except he's not attacking, he's trying to fend off an assault Ada isn't making. The stink of his fear is so loud in my nose that it actually blocks out all the rest of it. For this moment I'm locked on every twitch of him, and the rest of the world doesn't matter.

(*Good Dog* goes that part of my head that I shouldn't even have any more, the echo of heredity.)

Irae's channel: *Nearly shot him then.*

My channel: *That would be bad.*

Not mentioning how I'd been about to leap out and shake him in my teeth.

Ada's channel: *Put the gun away, please, Irae.*

Irae's channel: *You know me, Ada. Calmness personified. No itchy trigger fingers here.*

Irae's gun is shoulder-mounted and needs no finger on the trigger.

Ada's channel: *I know you when you're hopped up, Irae.*

Irae's channel: (Snickering.) *I'm cool, Ada.*

Ada's channel: *Stay cool.*

And then she's turned her attention to our guest.

He's frozen, regarding her with horror, with terror. And he looks weird to me, because I'm used to humans being a particular way – skin, face, hair, eyes. You can see a dozen different ancestries amongst the people back home, but it's all written over with what they did to people to acclimatise them. What we're still doing to people, in measured stages over their first twelve years. What we're doing to all of us, to all the various extents that our Bioform model plans allow. You see, now, why we couldn't just give in to prepper paranoia and chaos, up there. We all lived or nobody did. Not like here on Earth.

So he looks weird to me, but I have perspective. I know the meaning of that weirdness – not even a presence, but an *absence* of changes made. But Ada plainly looks very weird to him and he doesn't know what it means. I guess Tecumo's body probably freaked him out too, but not enough to stop him stealing from it.

"Hi," Ada says. Her voice comes out weirdly loud. It makes you thankful for the channels, honestly. In this atmosphere we're all of us still learning how to speak. It's an effort, but it's like trying to force a door that isn't locked, you end up shouting when you don't mean to.

The local cringes at her shout, and she grimaces – probably it looks like a furious scowl to him – and says the word again, more measured. When he doesn't greet her in return, she tells him, "My name is Ada Risa. What's your name?" Speaking like he's stupid or deaf, or he's a child and she's not good at talking to children.

Still, it seems to work. He stops trying to dig through the concrete with his shoulders. His wide eyes are still even wider with alarm, but just giving him something simple like a name has helped him. The stink of his fear ages, falling into the past. His new smell is still panicky, but I can tell he's more under control. Less likely to do something dangerous because he feels under threat. For now.

In the lull the chaos of the world threatens to break back in on me. All that myriad of individual and competing cues. It's not like that back home. There are the machine sounds and the people sounds but there isn't the *world* sound. All home has is weather. Earth has *life* and nobody ever warned me how riotous every little piece of it is to the ear and the eye and the nose. Which is why I failed Tecumo. Which is why I must *not* fail Ada. I force it out. I focus. The man is speaking.

His voice is thin and high. "Cricket," he says. Which is a name, maybe. Also an insect. Also a sport. Also, it seems, a name. And then he says, "I warn you, I am a monk of the Apiary, and I don't know what you are but you should know, if you do me harm, I am protected. There will be consequences." More of that piping voice, and I think that he's putting it on a bit, or that it's not quite the natural way he speaks. A sense of forced squeakiness to him. Unless that's just the fear.

Ada digests this. She's sitting down, cross-legged, leaning back from him.

Irae's channel: *Something personal? A weapon? Implants?* (A momentary image taken from her viewpoint: the line of her gun targeting this Cricket.)

Ada's channel: *Put that away.*

Irae's channel: *Cool, remember. Cool, but ready to turn up the heat.*

Ada's channel: *I think he just means his people, his in-group. Probably this outfit he wears identifies him as under their protection. Do not shoot him.*

She leans forwards a bit. "Tell us about our friend, Cricket."

8

CRICKET

Not really any confusion over who the monster woman meant, given that she and the corpse could have been close cousins. This woman had the same eyes, nose, pallor. And yet there was plenty different beyond that, as though these were just elements of a disguise slapped over a more regular face.

His mind scrabbled at the fact as he stared at her – overplaying the terror now so he could win some thinking time. *That means... there are a lot of them? They are like us, a melting pot.* Which suggested that somewhere there were, what, thousands of pallid little people like this. Or she and the dead man were like regular people from the Old, when people from across the world could travel and mix freely.

A thought, and because he was so tense it became words immediately. "Are you from underground?"

She blinked. "What?" And actually he didn't really hear the word in the nasal, barked utterance, just the sense of it. He'd barely heard her previous statement as actual language, because she had a weird accent, and because her voice was droning and distorted like bad audio playback from an ancient file. The interrogative jut of her query made the intent plain, though. She was human enough for that.

"Underground," he said. "A bunker, underground." It

had been a story, about when the Old fell. That some people – cults, corporations, governments maybe – had gone deep. Built themselves whole cities beneath the earth to hoard their stuff and wait out the end. Nothing in the Apiary records went to the truth or otherwise of it, because the end of the Old was when record-keeping began to fall apart. The failure of the global information network, a deluge of artificially generated false testimony. Nothing from a whole decade could be trusted, and even that nothing was rusted full of holes. All that was left was increasingly embroidered folk tradition, three generations on.

The woman said something, and this time he didn't even catch it, just shaking his head and frowning at her, mumming a hand behind his ear to indicate his blankness.

"Mole-people," she said, slowly and probably as clearly as she was capable of. And there had been a bit more, gabbled at the start, and it could have been "Are we?" and it could have been "We are." He didn't know if she was admitting to it or not.

He thought of small eyes closed against the grit of digging, and shivered.

Then he thought of the monsters, the long-limbed hairy thing that had grabbed him. The Old had a lot of different shapes of Bioform, not just the dogs the Factory made. Had they ever made men out of moles? People-moles, to live alongside these mole-people?

"I'm sorry about your friend!" he burst out, because her first question was still hanging in the air between them. In the dank backwaters of his imagination was the image of some hairy monster erupting from the earth to drag him down into the stifling depths. Great gnawing teeth. Huge, stone-rending claws.

She wasn't looking at him, for a moment. He realised,

ordering the last few minutes in his head, that it wasn't the first time. A glance, off to the side, away from the Cache shelter. Her monster, hidden.

Monsters, plural, he reminded himself. Or he thought so. Unless the thing that he'd felt stalking him all the way from the corpse had been his imagination, or had got in front of him...

He pulled his hood up. The spire of it was a bit bent, but it was better than a bare head for evoking the dreadful mystery of his position. "I warn you," he told the pale creature, making his words as forceful and precise as he could through the quaver. "I am an envoy of the Apiary. You may not know what that is, foreigner that you are, but perhaps you preserve records of the Great Bees of the Old. Know then, that we, who preserve so much of the past, do so in their shadow. Those who befriend us know us as even-handed with our blessings. Those who threaten us know that they shall never be safe from the Wrath of Bees."

And what would mole-people care about Bees? He wasn't even sure how much regular people believed that there really was a Bees. Even the old horror stories from before the fall of the Old weren't told so much these days. Bees, the enemy of humanity, the thinking hive.

An odd thought, then. A recollection, from years back, from his reading. All the monks were encouraged to trawl the archives in their free time back at the monastery. Knowledge left untouched was worthless. And every monk wanted to read about their patron, Bees. And Bees had gone to...

For a moment he felt himself on the cusp of revelation: what this goblin creature was, where it might have come from, but the details slipped out of his head when the woman leant forwards and jabbed a thin finger at him.

"Apiary," she pronounced, "is another word for factory?" She was still just sitting across from him. Had barely changed position in fact.

"What? No." And perhaps they knew he'd come from there, although that would suggest they'd been watching him even before he found the body. "Apiary. Where Bees lives. Not Factory, where Dogs live." Speaking weirdly broken language in an attempt to be more comprehensible, because maybe she was having the same difficulty with his words.

She seemed very angry, but at the same time very tired. She sat back again, and he saw the shift under her skin. The rods and bolts that the dead goblin had, moving with her. *Like extra bones.*

"I did not kill your friend," he said very slowly.

In reply she brought out his own picture recorder and showed him it. She'd unlocked it somehow. It was displaying one of the images he'd preserved, of the corpse. "Your work," she said.

"No!" His voice rising because, again, she'd glanced away, as though checking something was still there. Only, in a different direction this time. "Or, yes, the image. Not the death." It was just like on the road. He could see nothing of the monsters, other than the little goblin creature before him, but they felt closer and closer.

"You say you are from Bees."

"Yes, and Bees will harm you if you harm me!" he practically shouted at her. "Bees sees all, gets everywhere! Everything you do will be marred. You will live a life of misfortune and every piece of ill luck that lands on you will be the work of Bees." And he realised he'd started to gabble again and that probably meant she hadn't understood him. And while the standard Apiary threats maybe impressed most villagers and Bunker soldiers at least a little, he didn't think they were going to work on mole-people or... or... The gap in his memory was maddening. Because he was sharp as a tack for a few days but, after that, recollections decayed rapidly unless he wrote things down.

She frowned. Another punctuation to whatever they were having in lieu of a conversation, shot full of odd holes, as though she needed lots of thinking time for even a simple sentence.

As though she was talking to people he couldn't see, in a way he couldn't hear. And cued by that, he saw the minute motions of her jaw and throat. And that was another thing he knew, from the records of the Old. Probably nobody these days had an implanted radio, but it had been common military tech, and standard for the way most Bioforms had been built.

His breathing was no longer leaping at the cage of his self-control, trying to get away from him. His heart still pounded, but less like a bomb about to go off. *First, understand*, as Prior Stick said. Learn, study act. The monk's way. Make no assumptions. Believe no tales without consideration or proof.

The Dog Factory was an oasis of Old technology, where they could build Bioforms still. A retention that had made them an inviolable bastion of power in the Arreno River region. In living memory they were the only such holdout Cricket knew of. The Bunkers and the Apiary did their best with what had survived the fall, but it wasn't as though the monks would be building their own monster dogs any time soon.

But the world was larger than most people remembered. Of course there were other places, with their own traditions and domesticated monsters.

BUSINESS

The woman shifted, painfully it seemed, her silent conference done. "What happened to our friend?"

"I do not know," Cricket said back to her, trying to mimic the baby-talk rhythm of her words. "He had been shot. Probably the Bunkers did it." And he came close to telling her that she should come with him – sans monsters preferably – to the Apiary, to consult with Bees and Prior Stick. Except what if that turned out to be a terrible idea, and his fault? Better to get himself back to the monastery and let Stick make the call on whether to invite these dangerous new players in. He settled on, "Let me go in peace. You do not want Bees as an enemy." Because the one thing he was sure of was that Bees was a name to conjure with, for the mole-goblin-people. Even on those alien features he'd seen recognition of the name.

"Take us to your Bees," the woman demanded, and then actually winced, seeming to find her own words embarrassing. "To your leader," she said. "Take us to your leader."

9

WELLS

Irae's channel: *Take us to your leader, seriously?*
Ada's channel: *Do not start with me.*
Irae's channel: *You've seen the old recordings—*
Ada's channel: *I'm warning you—*
Irae's channel: *Of Martians—*
My channel: *Irae, enough.*

It subsides, snickering. I can see the little thief shifting, maybe wondering if he can make a run for it. Or… no, he's sneaking glances past the edge of his hood. Looking for Irae and me. Looking mostly in the right direction, too. A sharp tool, not just the timorous idiot he seems. Sharp enough for killing, surely.

He shifts his shoulders against the slab and speaks. If I cock my ears right I can hear him perfectly well, and I'm getting used to his weird way of speaking, the slurs and dropped consonants. The language he's speaking should be close enough to one of our more common dialects back home. We hadn't really anticipated any major communication difficulties. After all, there were several very separate polities with this common language back in history and they had no difficulty maintaining relations for centuries. Of course that was with day to day contact. We've been out of touch with these people for – what, four or five generations? And in that time the drift of speech patterns means I really have to force myself to concentrate entirely on the words. And that's just this region. With the limited communication and travel prospects for these people, probably every few miles has its own variant.

It all seemed so easy when we were planning for this back home. Tecumo made it seem easy. Earth! What was there to be worried about? We'd come, we'd help, they'd love us.

I think we are all feeling unloved at the moment. The three of us who remain.

The thief has said he needs to go speak to his leader to get permission or arrange a meeting, something like that. He's getting more and more jittery, and the background stink of his fear doesn't really tell me if it's just his awareness of us, the hidden watchers, or if he's about to explode into frantic action. So maybe it's true and maybe it's not.

Irae, needless to say, thinks lie. Irae says that it's running cool but I think its thermometer is broken because it sounds like it's on the very verge of going what it calls 'Goblin Mode'. We all have our problems, the three of us. We're all having our own separate difficulties acclimatising to what should be our natural home.

Ada's channel: *I'm sorry, I can't just trust him to go off and do anything else but warn people against us.*

Irae's channel: *You mean kill him?*

Ada's channel: *I do* not *mean kill him. Gun away, Irae. I'm serious.*

She has no way to enforce it. Just words. Irae's gun is still out. I can't see it but I'm absolutely sure.

Still.

My channel: *We cannot let him go unescorted. Let me come out. Tell him I'll go with him.*

Ada's channel: *No. You'll terrify him.*

I think the thief could do with some terror, but I'm not

Irae, to ignore Ada's wishes. The old dog in me, the pre-Bioforming one, knows its place.

"It's like this," Ada tells the thief. "I am an ambassador," making sure she says the words clearly, but trying not to sound patronising. Seeing he understands the word, she nods. "We have come from..." And I wonder what she's going to say. Underground? Pretend to be the mole-people.

But then the thief bursts out, "Mars!"

Ada stares.

"I remember now. There were people on Mars. Another planet." In case, apparently, we don't know what Mars is, despite coming from there. "In the Old," he adds. "People and – Bioforms." And here's this savage done up like a wizard and suddenly he's talking science at us. "Am I right?" he asks. "Is that you?"

And we'd had the discussion, obviously, and we weren't going to pretend to be locals for the simple reason it couldn't possibly have worked. All our mummery so far was because we'd just assumed that this ignorant little primate couldn't get his head around who we were. Except it's not been *that* long since the communications crash, and except...

"What sort of a place is it, that you come from?" Ada asks him carefully.

"The Apiary," he says. "We husband the learning of the Old." And this time I catch the implied proper noun in the way he says the word, and understand it's a noun in its own right and not just a lost adjective. "We have records about you. Your... ancestors? There are people still on Mars?" His eyes, already huge, are growing wider and wider. "You still have the tech to cross between planets? You've not..." *Lost it all*, are the words that tail off into silence, I think. Revealing the gaps in his understanding, because if we'd lost that much then we'd not be here. And also that it's easier to get into space from Mars than it was from Earth.

"All right, Mister Cricket," Ada says heavily. "This is how it's going to be."

"Siblen," he says.

She frowns at him.

"I'm a monk. You should say, 'Siblen'." And it's like he's lecturing, like he's forgotten he's afraid of her, and of us, her two unseen friends. All that twitchy suspicion has dropped from him. He's almost eager.

"All right… Siblen Cricket," Ada says, "you and me are going to your monastery. To meet Bees. We know Bees. Bees is our friend."

I swear if his eyes could get any wider they'd pop out of his head.

Ada draws in a big breath to explain the problem.

My channel: *Don't. He's not a friend. He'll use it against you.*

Ada's channel: *What are the options here, Wells? Someone has to trust first. He seems to have calmed down. They're scholars of some sort. That gives us common ground. And Bees!*

And she says, "Only I'm going to need your help to get there, okay?"

He doesn't get it, frowns at her. She gestures helplessly at herself. Literally helplessly.

"I can't walk, Cricket. I've had a… medical problem. The gravity's hit me harder than we thought, and I just… tire out. So when I say 'take me to your leader', I really do mean it. You'll have to take me."

I can't tell if he knows 'gravity' or not. He frowns again, looks at his hands. "I don't think I could carry you, Ma'am Adarisa."

"What about your wagon?" Ada presses. "My – I was told you had a wagon."

"I... left it behind when I was running away from your... Bioform."

Irae's channel: *Quite the little diplomat.*

And, from over the lip of the depression they're in, comes the wagon. I can just about catch a glimpse of Irae as it shoves the conveyance forwards until it begins to roll down. A shimmer and a shift, a moment when my eyes catch the long, low shape. And then I lose it again.

The thief has seen nothing of Irae, just the wagon appearing of its own accord and then trundling down the slope. Hitting a rock and turning onto its side. It's a makeshift little thing but Ada's small. She'd fit in it.

My nose is full of his fear spiking again. The haunted wagon has got to him. The reminder of *us*. Those nervous twitches are coming back. I force my attention on him, every little move, waiting for the moment he turns out to be dangerous after all.

Ada, sensing a little of the same, offers him a tin. "Would you like some food?" It had been the original plan, derailed until now. Bribe the locals with what is, really, their own goods. We didn't think he'd trust our rations, and they probably wouldn't be very good for him.

He looks at the tin, and then his eyes stray to the earth where we'd dug it up.

"You raided the cache," he says. His tone suggests this is a bad thing. "That's a Bunker cache, a Griffin cache. You have to put it all back. You don't steal from them. They'll hunt you down."

"You said 'Bunkers' before." Ada seizes on the word. "When you said about Tecumo – our friend, who died."

"He got shot. Maybe it was the Bunker. Probably

the Griffins. They're the big one near here. The Factory said there were Bunker raiders nearby. They don't need much excuse." Seeing her not get it, he goes on urgently. "Soldiers. From the Old. They don't make or build like the Factory, or preserve, like us, but they have a lot of stuff held back from before. Guns mostly. They – you can get on with them but they don't take insults, theft, anything. You can't push them."

Irae's channel: *Try me.*

And, as if that's all the world has been waiting for, the shooting starts.

I've done it again. It happened to me again. The world, its overwhelming rush. Just two shots, and pitched high to strike dust off the concrete over Ada's head, but I didn't know. Didn't hear, didn't smell, didn't see. Failed. I failed.

I bunch, take it in, broaden my senses. I'd screwed everything down to focus on the least little twitch of threat from this Cricket, and all the time he wasn't the enemy. The enemy was this Bunker who killed Tecumo. And now probably they've found us.

They're coming in fast. Two vehicles, actual vehicles. Electric, near noiseless, but if I'd had my ears up I'd have heard the crunch of tyres and the hum of their motors. If I'd had my nose to the wind I'd have caught the scents of metal and gun oil and sweat. You can't surprise me, that's been the boast of my model for generations. Our senses are second to none, millions of receptors and decoder neurons all wired into our enhanced Bioform brains.

And two whole cars crept up on us and I didn't know.

It's this world. It's this Earth. Mars has life now – the bioengineered scurf matting that turns dust into soil, the agricultural caverns and domes, and us. But our home-world is sterile desert still, even now atmospheric pressure and oxygen are creeping upwards across the red globe. Engineering organisms that can disperse and grow is still a work in progress. So of course I can hear a pin drop and smell the slightest lie, when I'm on Mars. And I never – we none of us never – thought about what that would do to me back on the old ancestral turf.

I go from tight focus on the monk, shutting everything out by sheer force of will, to trying to take in the whole situation in one go. It's like taking a bat between the eyes and into the centre of the brain.

My channel: (No words just howling.)

Ada's channel: *Calm, Wells. Stay back. Stay hidden. Those were just warning shots. This doesn't have to turn into—*

Irae's channel: *Who are these clowns? Local Max Appreciation society?*

Irae hasn't shot them, I register. Meaning that actually it really was running very cool indeed, reflexes measured in whole seconds.

Ada's channel: *Gun down, calm down, don't show yourselves. Cricket says these are the Bunker people, whoever they are. Some kind of militia. A major polity in these parts, I guess. I do not want our first meeting with them to be a bloodbath.*

The cars are drawing up. One is closed and armoured – and then re-armoured by someone inexpertly adding plates to it, so that it cants to the left where the weight isn't distributed properly. The other is open, faster, just a cage of bars over the top for all the people with guns to hold on to. And there are eight of them visible, plus whoever's in the closed car, at least two probably because there's a turret. Supposed to be some sort of ordnance there but now just a slit and the barrel of a rifle thrust through it.

The men are not what I would have expected, if you'd said 'gun-toting militia' to me. Yes, they have some effective-looking anti-firearm body armour that absolutely isn't going to help them against the ammunition Irae's packing. Beyond that they've got cloaks and they wear their hair long and tied back. There are rags and strips of coloured cloth tied here and there about them, giving them a feel of carnival I suspect is not warranted. All of them are men, with drooping moustaches and bare chins. Plenty of tattoos: all the usual symbols of human violence through the ages. Skulls, darts, daggers. I see some spread-winged icons which speak to some very dark recurrent political affiliations in Earth history. A second squint at the bigger decal on the closed vehicle shows not the expected eagle, though, but something with more diverse parts. A maned, beaked head in the centre, and a marching ring of talons and clawed paws and tufted tail around it, flanked by the two spread wings. So: nuanced, but not necessarily any better.

Irae has been thinking along similar lines now it's cranked its metabolism up. *Bad guys. Gonna shoot 'em.*

Ada's channel: *You are not. Hold fire. Let me talk.*

Four of the Bunkermen jump down from the open car straight away. They are armed with robust but old-looking guns that have been painted bold colours, bound with strips of cloth like their clothes. They have the same rags about the ring pommels of their long knives. Swords, almost, metre-long blades in fancy stitched scabbards. A group amongst whom *show* is very important. The term comes to me: *honour culture*. Also bad news, because it means they'll do stupid, harmful things to their own detriment if it's that or lose face.

The four swagger in. I can feel the high whining building up in me because there's only so long I can keep focus without either exploding into action or losing track. A thousand thousand living things are all around me. The air is full of their sounds and scents and the fleeting trace of motion. Every plant is an unfamiliar assault on the nose. Every insect keens in my ears. And these men, these angry, armed men, are the threat, but because they're further from me and not shouting, they almost fade into the background. A riotous assault on every sense that a mere human couldn't possibly know. They had dog-forms at the Factory! How do they cope?

"Sibbo!" calls one. His voice has an accent that's three-quarters the same as the little thief he's addressing. "Monks know better than to steal from us, do they not?"

He pauses, the others fanning out behind him in a way that makes plain the speaker is the ranking leader here.

Cricket stands up. He's got his tall hood up, and his hands out. The layers of robe he's dressed up in do a good job of obscuring almost every parameter of him. Anonymising and dehumanising him. He could be either gender within a range of thirty years.

"Greetings to the Griffin." A vague gesture, maybe benediction. There's an interesting shift amongst the soldiers. Right then it almost has me snapping, bounding out and laying into them. A moment later I interpret it as unease. This Apiary carries some weight with them, and no surprise given who Cricket said was running it.

"Monk," says the Bunkerman leader – and that wheel-and-wings design of them has enough lion in it to be a griffin. Just an imperialist eagle with more claws, though. "Monk, you are within our land and you have not come to pay your respects."

And probably Cricket hadn't been in their lands before we grabbed him and brought him here. Not knowing the local taboos we were breaking, or how anything works here. We were so very confident we could work it out. We, the superior, the advanced culture come home to see our backwards cousins!

The whine is building inside me. The world is building, outside me. A terror of getting it wrong again, failing Ada, that will drive me into getting it wrong and failing Ada. The wrongness of action, the wrongness of inaction, the vice of making decisions.

Irae's channel: *Target acquired.* (And an image of the lead Bunkerman's centre parting, high magnification.)

Ada's channel: *Irae, no!*

And then she's been spotted. Slumped in the shadow of the concrete, she'd gone this far without drawing attention. Cricket, to his credit, hadn't pointed her out to his fellow unaugmented humans, either. But the Griffins have been casting about like good soldiers – alert, like I should have been but have lost the capacity to be.

"You, get up, get out into the open," says the leader. Of course he does.

"I," Ada says, and no dog ears needed to hear the uncertainty, "can't."

"She's sick," says Cricket. "I was tending her." And I'm impressed because plainly he's ad-libbing but it comes out from beneath the lip of that hood with some real authority.

"Go get her out here." The leader isn't having it. Two of his people go for Ada. I want to watch but there's sound and movement at the car that drags my attention over. It's nothing. It's just one of the other Griffins out with some tools and taking the moment to do something under the hood of the car, because there might be a showdown with a monk but someone takes their duties seriously. Like I should, but can't.

I miss the moment when they actually see Ada properly, but their startled yell drags me back to them. The pair have recoiled from her, one has a gun levelled, practically right in her face. Cricket has his hands up and apparently he's decided he's on our side against his neighbours, which says a lot about politics over here. He's calling for calm, in a forced-sounding sonorous voice he's putting on. He's invoking some sort of benefit of clergy. He's saying Ada is under the protection of Bees. The leader of the Bunkermen is shouting at his subordinates and they're shouting – more respectfully – back at him that they don't want to touch this thing they've found. And it's all too long and it's all too much and booming in on me, the words, the anger, the smell of them – rancid sweat, alcohol, machine oil and some sour stuff in their hair. I'm whining now, deep in my chest, and I can't stop it and Irae was supposed to be the irresponsible one I'm the good one I know my duty they brought me along to protect them to do the right thing but Tecumo is dead and Ada and Ada and Ada—

10

CRICKET

"The protection of Bees!" Cricket was declaiming. And it was working to the extent that they hadn't knocked him down or laid hands on him yet, but it wasn't stopping them going for the Martian. And she was calling too, in her twist-accented nasal voice.

"I am an ambassador! Please!" And she actually hauled herself up on shaking legs, leaning against the concrete. He could see her lead-and-chalk skin sheen with sweat as she gasped for breath, hands out to forestall them. "Please do not lay hands on me." And then, "No, Irae! I forbid it! Not like this!"

It was so unexpected the soldiers stopped, because it obviously wasn't directed at them. Cricket understood. He cringed inside. His own shouting was keeping a little space about him clear of fists and gun butts but it wasn't transferrable to the alien woman.

One of the soldiers drew his sword, more out of something like superstition than threat. A lot of things from the Old could have you sounding mad like that, and some of them were catching. And many of those Old things had survived the fall in some way more substantial than stories.

"Irae, no!" Leaning, shaking, that horrible inhuman face twisted. "We didn't come here to – Wells? Wells? *Wells?*"

The man without his sword out went for her, yanked

her arm, hauled her out into full sight of his commander. She fell, ended up with her arm twisted at a painful angle. She was wheezing frantically, clutching at her chest, but her face stayed dead white, none of the red or purple he'd have looked for.

That was when it happened. Or, rather, it had begun to happen a second before but Cricket hadn't noticed. None of them had.

The monster broke cover. The one he'd known was there but, because he hadn't actually *seen* it, somehow hadn't quite believed in. The monster that had seized him at the gates of Brokebridge. And perhaps it was supposed to be a dog, but the Factory would not own it. It was no more a dog than Ada was human.

It was wearing the same kind of plasticky outfit she was, or the corpse had been, but tailored for a shape that had very little to do with the human. Cricket knew Bioform dogs. They were compact, muscular. They stood like humans more than they went on all fours. You could see the wrenching and the changing that had been done, to the way a dog's skeleton was born to be, to make them more fit to the role of talking, thinking creature. Someone had done that to a dog, in the history of this *thing*, and then done a whole new series of terrible things to the thing that dog had been made into, and made a monster.

It was big, but without the chunky solidity of the Factory dogs. Its outfit ended long before its limbs did: long, spidery, shaggy with reddish hair. It was hunched, arms longer than its legs and tearing up the ground on hands and... feet that were also hands. Its face was less dog than some carrion thing, a melange of other influences hacked into it, and mostly jaws and huge ears, the eyes almost lost in creased flesh and hair. It was like someone had de-winged a monstrous bat and set it to scuttle over the ground at twice the speed of a sprinter. It could have kept pace with the cars.

One of the Bunkermen was fast enough, seeing and turning, to get a brief rattle of automatic fire off. The shots didn't impact, and then the monster had caught the man with its claws and *flung*. Just scooped and thrown. Cricket saw things ratchet and move under that hairy orange hide, a whole extra skeleton like the human Martian had, but in full working order. The gunman was gone from his sight so suddenly it was like a conjuring trick, save for the arcing shriek overhead as the loose body spun end over end before hitting the ground in a crunch of broken bones.

The other soldiers were reacting now, the commander's shouts barely needed. Guns tracked, but the monster – *Wells?* – was amongst them before they could make use of them. Those appalling jaws, the whole front end of the beast, closed on one man and shook his joints and vertebrae loose in an appalling economy of motion, then used his flailing limbs to batter at his friends.

Someone from the armoured car was yelling for them to get clear, give the turret gunner a shot. Honestly when the Bunkermen scattered it was probably more to get away from the monster than for tactical considerations. It caught one of them by the heels even so, lashed him across the ground like a whip.

The turret gunner opened up and the monster moved, that hideous long-limbed scuttle keeping it ahead of the puffs of earth where the bullets scarred the ground. Cricket was backing away now. He should go to Ada, surely. He should be a good monk and hold to his responsibilities. But going to Ada would bring him within the reach and the ire of that beast, and suddenly he didn't want to be the man who introduced the Martians to Bees. He didn't want to be their friends. Because they were monsters.

There was a deep metal sound, and he saw a hole punched through the armour plate of the turret. Even as he registered it, there was a sense of movement too swift for the eyes, a new crescent bitten into the turret's viewslit, and the gun within fell silent.

There were no Bunkermen near Ada any more. No living ones at any rate. They were fleeing for the cars, while others laid down a wild scatter of fire to try and ward the beast off. Cricket saw a man just punched straight from the open car, his chest staved in by the force of the shot. Not from the dog-beast; from the long grass.

The open car reversed violently, and then the driver's head exploded in bloody shrapnel. Ada was shouting "Irae, no! Irae, stop!" But there was no stopping it and no undoing the damage. Cricket was watching his world torn apart. That it was a particularly brutal and vicious part of his world didn't matter. It was his, and these terrible creatures were not.

He ran. He very nearly ran straight into a bullet because the Bunker soldiers weren't in a mood to be discriminating. Somehow, though, he didn't get shot. He reached the armoured car, just as the remaining soldiers from the other vehicle were abandoning their ride to clutch at straps and handles on the one transport that still had a living driver. Cricket waved his arms at them wildly. His enemies, the greatest current problem the Apiary had to deal with, but he'd take his chances with them over the monsters from Mars.

"Don't leave me!" he cried, monastic dignity forgotten. Practically hurling himself into their blades and gun barrels, until one of them hauled him up unceremoniously by his belt. *Saved!* he thought, and then the tight clench of his collar was choking him as the man twisted it. The cold circle at his temple was the barrel of a gun.

"Stay back!" the Bunkerman yelled out at the monster, at the world. "We have him! We'll shoot him, you come one step closer!"

The armoured car reversed violently into the grass, spun into a skewing turn and took off. Every jolt seemed an open invitation to the man's trigger finger, or else for a shot to come from nowhere that wouldn't even leave a coherent corpse to deliver to the monastery gates. But perhaps Ada's shouting had restrained the unseen sniper, because no more death came from nowhere to bedevil the retreat.

The strangling noose of his own collar didn't loosen. The look on the face of the man who had him didn't get any less murderous. He had one elbow hooked through a canvas loop bolted to the side of the car, and the only concession he made with the gun held in that hand was to shift it from temple to the hollow of Cricket's throat as he stared back the way they'd come.

11

ADA

It had all been so simple on the red planet. A humanitarian mission, from the benevolent survivors of the Crash to those who had crashed. A voice out of the old homeworld begging for their aid. A last holdout of that part of Earth which had dreamt of Mars and space travel and a bright future. And of course they had to go. Ada and Tecumo and their fellows hadn't really even considered the alternative: to just keep listening to that radio channel until it wasn't broadcasting any more. How tragic would that be?

There had been uncrewed sorties past Earth before. One work crew or another had put together the resources to launch a probe out of the weak Martian gravity and send it to buzz around Earth a bit. Taken orbital images of the ground there, looked for power usage, large communities. And there was power usage, and there were large communities. If, by 'large' you meant one per cent or less of what had been. The great cities of Earth lay vacant in some parts, turned over and haphazardly converted to agriculture elsewhere. The concentrations of human life they had once supported had died or dispersed because that density of life was a hothouse flower absolutely reliant on a vast and complex transport network. It all came down to food, fuel, medicines. Things that, as the global system of inter-reliance and trade buckled and collapsed, could not be manufactured in sufficient quantities, nor carried from where they were made to where they were needed. Oases survived for a while, The whole first post-Crash generation perhaps, while Mars endured its own knock-on turmoil and couldn't have helped. But the sophistication of Earth required a million million parts and pieces and resources, spread out across that whole Earth. A fist of worsening climate, petty border wars, power-grabs by the brutish and greedy, ideological platforms of the ignorant and fanatical, it had all clenched down until the great underlying armature of human society became too bent out of shape to support that level of civilisation. No one great natural disaster, no all-consuming world war, no catastrophic pandemic. Scores of storms and freezes, droughts and floods. Dozens of vicious, selfish regional conflicts that only destroyed what could no longer be rebuilt. A half-dozen major disease outbreaks, born of shifting climate bands, human movement and regressive medical policies. Not one finishing stroke for the great global human society, but you could still bleed to death from a thousand cuts.

And Mars persevered. Because of its oldest inhabitant.

When the people of Earth had decided to colonise Mars they had sent Bees. The science of Bioforms was established by then. Animals engineered genetically and cybernetically into war machines, at first. Later, in more enlightened times, just into people who could do things humans couldn't. Always a source of tension, but there had, Ada knew, been a golden time when humans and their Bioform creations had been building something great. So she'd read, so she believed.

Bees had been an early part of the Bioform project. An insect swarm with an intellect spread across its parts. One of the first distributed intelligences and, because Bees never went away and never stopped improving herself, still the greatest of them. When the Mars colonisation began, Bees had been the great hope of the world, capable of finding solutions to problems that humans and Bioforms hadn't even thought of. Send Bees to Mars, Bees would survive, start building, prepare the ground for more conventional colonists.

Then the anti-Bioform backlash began on Earth, because most Bioforms were bigger, stronger, more capable than baseline humans, while being far less numerous. An easy target to make people afraid of. And if the anti-Bioform backlash was bad, the anti-Distributed Intelligence crusade was far worse. DisInt became the devil, to a great many human demagogues, and Bees hadn't helped her case right then. She had run out of patience with human prevarication and tried to unilaterally alter the way the world's energy economy worked. Nobody wanted to have control of their power stations just wrested from their hands at the whim of a swarm of insects.

Bees became the enemy, on Earth and on Mars. Bees went to ground, and all the other DisInt networks were imprisoned, broken up, exterminated or driven into hiding. The other insect nests, the Bioform packs, and most especially the human Distributed Intelligences who became a terrifying fifth column for anyone who wanted to whip up a mob and win a political office.

On Earth that was how it stayed. On Mars, not so much. Mars and Earth never had a grand falling out, exactly, but after some ructions involving Earth politics, Mars ended up at least semi-independent, through the assistance of Bees. And although Bees then crawled back into the dust and rock of the red planet – to concentrate on her own high-minded schemes that had nothing to do with human endeavours at all – the powers of Earth never got round to strangling Mars or invading it or levelling sanctions. Because the threat of serious Bee-related retaliation remained.

After which things got worse. First as single spies, as the saying went, then in battalions. And Mars braced itself, and worked to become as self-sufficient as possible because it was obvious that was going to be the only option. Mars pulled together. Earth fell apart.

A few generations passed, during which the embattled population of Mars – the engineered humans, the animal Bioforms – at first clung on with their fingernails, and then got back onto their feet. It hadn't been inevitable. There had been a time when the strains of trying to survive alone had looked to be cracking all the domes across, with the atmosphere outside not yet thick enough to support life. That was when Bees had come out of the red desert like a prophet and brought their incomparable ability to see a problem from a thousand different sides.

Mars survived, which meant holding on to absolutely all of its technological systems because without them the artificial farms would starve, the atmosphere plants would stall, and everyone would die.

Earth survived, for a given value of survival, but the great technological high wonder of its civilisation did not, because it didn't have to survive for humans to carry on.

Fast forward to the present day. A signal, a plea for help, a consortium of Martian work crews finding the resources for a mission. A triumphal return to the blue-green world of their ancestors.

It had all gone so horribly wrong.

Ada shifted awkwardly, feeling her breath rasp. Her lungs fighting the weight of her own body to drag air in and force it out. Her heart labouring. She was on four different medications right now, to keep it strong and steady, and still she felt as though she'd tip over into arrest at one more shock.

The armoured car was dust in the distance. The other vehicle, the open one with the now-crumpled cage of bars over it, lay on its side, abandoned.

Irae's channel: *Well, at least they left us transportation.*

Ada's channel: *I've had it with your jokes.*

Snapped, angry, without the restraint the situation required.

Irae's channel: *I wasn't joking. I don't think I'm as funny as you think I am.*

Ada shook. More internal convulsions. If she'd been an Earthling she'd be weeping, but there were no tears on the red planet. Her eyes had a battery of other anti-gritting systems and their tear ducts had been re-engineered into something more efficient. Every child had the surgery, after children began to be born on Mars. Long before the Crash, even. After Mars won a tenuous arm's length separation from Earth politics, nobody wanted their long-term population to be beholden to whatever corporate stooges or prisoner details the current World Senate Security Committee felt like sending over. The first Children of Mars, each one taken from its native state and run through a comprehensive battery of genetic and physiological modifications to make them fit for Mars. Even for the Mars of a couple of generations' work, where the atmosphere composition and pressure and the temperature were less than instantly fatal, at least in the basin of Hellas Planitia where the colony was based.

Generations, growing up on Mars, knowing only those pale Martian faces, flat, small-eyed, large-nosed. Which must seem like bizarre troll-people to an Earthling eye. But they were still Earth people, under it all. As though it was all just elaborate costuming. They could always go home.

Well, here was Ada, proof positive that it wasn't actually that easy.

They'd known there were going to be issues. She was taking pills to de-oxygenate her system, because she was used to getting by with worse air and a more efficient respiratory system. They'd worried about the stronger sunlight, too, but while it stabbed at her eyes, the radiation-reflecting layer in her skin seemed more than capable of coping without even getting a tan. That pallor she had wasn't an absence of pigment but a presence. The problem was the gravity.

She'd gone through the tests. She'd had the surgery. There was a whole extra endoskeleton implanted under her skin and alongside her regular bones. Not to make her some kind of superhuman, but to support her weight under the homeworld's pull, to aid her ribs in letting her breathe. They'd swung her about in the centrifuge back home, and pronounced her eminently suitable, but she hadn't actually been walking about under one full Earth G back then.

She could walk. After a handful of steps she would be fighting for breath, feeling the lactic burn, the trembling of her muscles. The mechanical skeleton was trying, but its movements didn't seem to mesh with her own, jabbing like hot pokers all about her limbs and torso. She was useless here.

And Wells, the dog-Bioform, whose long-limbed body had taken perfectly well to the augmentation, had problems of her own. She was trying to hide it, but Ada had worked out just what part of her biology wasn't adjusting to her new environment. Her keen senses, honed over a life of thin atmospheres and limited stimuli. Earth was driving her mad with its clamour.

Then there was Irae. Irae the cold-blooded. They'd nearly not taken the reptiliform on the mission. Conditions on Earth were still climatically variable. The storm systems bred by runaway climatic chaos were still prowling back and forth around the globe looking for victims. A sudden freeze here, a baking drought there. Why bring a Bioform that couldn't even regulate its own temperature? But they'd fit Irae with an electric vest and given it the controls. It could be as cold or hot as it wanted, meaning as slow or as fast, body and mind. Except of course Earth was far, far hotter than Mars, and an overheating Irae was a capricious thing.

That was the first thing that had gone wrong almost immediately, really. Their ship had let them down onto the surface of the homeworld and three of the four of them had almost immediately found it wasn't like home in ways that practically incapacitated them.

Only Tecumo had been able to face Earth with equanimity. And he had decided that was enough. They wouldn't just call the lander to put them back into orbit, and eventually arrive back at Mars as shamefaced failures. Eventually meaning when the two worlds' orbits actually coincided enough to make the trip viable. After all, Tecumo said, they were right near where the plea for help had come from. Tecumo would just walk over to the place and introduce himself, make the arrangements. They could help. They were the Martians, here to save the day. *We come in peace.*

They'd shot Tecumo. Not the Factory, the place he'd been to see, but apparently these Bunker people on the way back. She hadn't gained much of a sense of them from Cricket's words, but their appearance and actions hadn't left much room for doubt. Presumably it was the presence of these Bunkers that meant the Factory was in such dire need.

And Wells had been supposed to be watching over

Tecumo. But Tecumo hadn't made allowances for Wells' problems, and Wells had lost him in the riot of the world. And he'd died. And there was no combination of words that Ada could put together that would make Wells feel better about that.

And then they'd finally found a native they could talk to. Some kind of monk, who'd talked about Bees. Unexpected, unlooked-for. A monastery somewhere with Bees as its leader, preserving the knowledge of the old world. Not necessarily something Ada would have guessed Bees would be doing, but within the bounds of possibility. And given that Bees had remained on Earth despite all the persecution – as well as travelled to Mars, given the divisible nature of Distributed Intelligence – the Martians certainly came here expecting to meet with her. Not like *that*, not through some quasi-religious proxy. But she'd take it. She'd go to the monastery and find enlightenment through this new incarnation of the old insect intellect. Except they'd lost the monk. They had, in fact, given the monk over into the hands of murderers he'd plainly not been fond of.

Ada shuddered again, then fought herself back under her control. Breathed deep, trying to sync organic ribs and inorganic exoskeleton to make things easier. "Right," she said.

The other two were watching her. Wells loomed over her protectively, so she was practically within the cage of the dog-form's long limbs. And fair enough to Cricket, Wells probably wasn't what Earth people thought of as a dog, but the Martian surgery-factories had always been keen to over-adapt. Four prehensile feet, long limbs, big ears, they were all useful things on a low-gravity world with a thin atmosphere. And with Wells, the exoskeleton had *really* taken. They'd over-engineered it for fear her limbs would snap otherwise, and even under Earth's drag she moved like monkeys and spiders. Ada could understand why the monk had fled.

Irae was out there. If she squinted, she could just about see the tell-tale lines and shapes that were its long, low body. But it was still as a dead log and the same colour as the ground, and every time Ada looked away she had to search for it all over again.

"We screwed up," she said. It was inarguable enough that neither of the others tried to argue the case.

"We call the ship for pickup," Wells grated. Her voice was a saw-edged rasp. All of them were pitched far lower than back home, and their voices came out with an extra layer of nasal vibration. Talking subvocally on channels was clearer, and for the Bioforms more natural. She'd only had one implanted for the mission, though. Audible speech still came more naturally to her.

"We need to find this monastery," she said. "If anyone knows what's going on, it'll be Bees."

"The Factory—" Wells started.

Irae's channel: *Didn't do Tecumo any favours.*

Wells' channel: *He was killed on the way back.*

Irae's channel: (An image: a twisting serpent. Meaning, a shrug.) *I agree with Ada, for once. We know jack. Let's find out. Let me go hunting. I'll find more monks, bring them to you. Or just grab some people. Someone will know where the place is. Rough up enough locals, they'll spill.*

"We are *not*," Ada got out. "Not doing that. Who do you think we are?" Speaking loud to reach Irae. Losing her breath again and fighting her own body.

Over her head, Wells growled protectively,

Irae's channel: *We are people with a dead friend. We're owed. Let's collect.*

"We came here," Ada gasped, "in *peace*. To *help*. We've. Only. Made. Things worse. We are *not*. Going to just start kidnapping people." A resolution came to her, in the midst of all that. The only right thing to do, that might redress all the wrongs even a little. "We are going to save Cricket from the Bunkermen."

"No," from Wells.

Irae's channel: (Snickering.)

"I am going to this bunker. I am going to follow the vehicle. In that car, if we can get it working. I will talk to them. I will explain."

"Explain how we killed a half dozen of their people," Wells growled. And fair enough, the bodies were still right there and around them. "They didn't seem the type to accept that as a misunderstanding, Ada."

"No, I will not accept that," she said stubbornly.

Irae's channel: *We declared war on them.*

Ada waited for Wells to weigh-in on her side, but the dog-form's silence only supported Irae.

"They will try to kill you," Wells said eventually. "Or imprison you. Irae and I will be forced to kill more of them to prevent that. Nothing good can come of this, Ada. Irae's right. Let it locate another monk. *Not* kidnap, just find. These monks are better, for talking to. Bees is better. We did wrong. We made mistakes. We can't always make good on our mistakes."

"I won't accept that," Ada repeated, knowing that the words, speak them as determinedly as she would, didn't change the world.

Irae's channel: *Well, fine then.*

A beat. Ada tilted her head back to find Wells cocking an eye down at her. "Fine what?" she asked cautiously.

Irae's channel: *Fine and I'll go get that monk back then.*

If you absolutely have to have the same monk. I think any monk would do, but fine.

"Irae, wait. What do you mean?" Ada demanded. Waited, without response, then gritted her teeth and spoke into her jawbone mic as she'd been taught.

Ada's channel: *Irae, don't just go off.*

Irae's channel: *I can't hear you. I've just gone off.*

Ada's channel: *Irae, come back here!*

Irae's channel: *Going to get you a monk. The same monk. Cricket the monk. Think he'll be glad to see me?*

Ada's channel: *Irae!*

The shuddering was back and she sagged into Wells.

Irae's channel: *I will let you know if anything important happens.*

Ada's channel: *I forbid you to kill anyone.*

As though she was the leader. As though any of them were.

Irae's channel: *I will let you know when I kill anyone.*

Ada's channel: *You—* She lost the mode of speech for a moment, just spitting sounds into the void of her own mouth. *Please, Irae. We didn't come here to make enemies.*

Irae's channel: (Snickering.) *There's an old story, didn't you know, about how the oldest enemy of everyone is the creature that crawls on its belly.*

And then a silence, very definite, as of a closed channel.

12

THE WITCH

The people of Clearwater spot me while I'm gathering the fungus. Not necessarily a problem but I always prefer to pass by the occupied places of the world without notice, unless I have business with them. Some of my sisters have been driven away. Some have been harmed, a few killed in particularly bad times. Times demanding a scapegoat, or when some local tyrant had been honing their people to hate the outsider, in order to build their own influence. Clearwater was ruled by a council last I heard, and was well supplied by its namesake spring and some good reclaimed land. Still, there was no permanent resident there whom I could connect to, and my last news was a month old. A lot could happen in a month.

So they see me. Some boy out with the goats, early morning. Then a delegation of adults, ostensibly out hunting an errant animal but very plainly come to see the truth of it. And they see. It's just me. Just this solitary old woman. Not actually so old, but I dress for it, and my face is creased with dirt, and the fungus makes all our hair go grey before our time. Just the witch, basically. For good or ill, the witch. And some of them will have heard tales of cures and potions and wise words from one such as me, and others of curses and wicked retaliation for wrongs and slights. Either way they don't approach or hail me, but leave me to my business.

It's not as though they have a use for these fungi anyway. Put *these* in a stew and your belly would know it soon enough. Nothing here for the regular human stomach to metabolise.

This is because the particular fungus I gather was bioengineered back before the General Collapse, in the time the locals quaintly refer to as the Old. It was bioengineered by me. Me? Not quite me. Frustrating, really. Back when the world was so much more connected, I should have applied myself to the linguistics of it. Found a word to describe that liminal state that is both self and other. But let me say *me*, even though it was long ago and before I was born.

This fungus – it's a bit of a hothouse flower, to change kingdoms. Being an artificial creation still clinging on in an untended wild world, it requires certain conditions. Human conditions, mostly. Being something designed to have a very specific effect in certain human bodies, it needs trace levels of human-generated biochemical genetic runoff. Hence it tends to sprout up close to human settlements, which is inconvenient. Which results in harmless old mushroom gatherers like me being spotted by alarmed locals of a morning. And it's not even a mushroom. More a yellowish slimy tracery growing on dead wood. But I gather fungi indiscriminately as I go, partly because I have a taste for the edible ones, and the other sorts all have uses of one kind or another. And partly because if the locals ever understood that I and my kind were so utterly reliant on this one species then some future hater could go on a crusade to eradicate it. And, in doing so, ensure that there was an end to me.

And I'm not so ill-prepared. There are certain places in the world where I cultivate the strain. But I have seen how that goes, in the face of a crusade. I have had my places of birth and nurture broken open by those who hated me. I have been hunted. I have hidden. Like now, when I hide behind the guise of mad hermit women wandering alone in the woods.

And I am alone, right now. I, whose chief defining characteristic since I became what I am has been *never* to be alone. To be a part of something greater. The oldest Distributed Intelligence network. No, not Bees. Bees, for all the hype, is marginally my junior. The other one. The one that was hiding in plain sight from the start.

Distributed Intelligence is another hothouse flower. The product of Peak Tech humanity, like so many other things. And like all those other things, when the General Collapse rippled across the world, it became unsustainable by regular means. Only a few isolated towers remain, of the great castle of achievement we built, and each one maintained solely by extreme effort or artifice.

The Bioforms are mostly gone, because to make even one was a great technological triumph. There are three places across that part of the world I have access to, where the science has been maintained. The place they call the Dog Factory is the closest, and even there I see signs of degradation. The dogs aren't what they used to be and neither is the Factory. I steer well clear of it. No gathering mushrooms in *its* shadow.

I, too, should be gone. But I had already been applying myself to unassisted survival long before the General Collapse. Because I had gone underground. Because they would have eradicated me if they could. Because DisInt, as it was called, had been blamed for all the world's ills, after the big fight with Bees, the Mars schism, all that. Those of us left on Earth were branded the enemy. Because, unlike Bees, we were a quarry they could hunt without much consequence.

I cannot, now, implant all the complex electronic business

that I once relied on. I cannot manufacture the thousand thousand little components, that would turn a human mind into a part of a great network of thought. Just as I cannot, single-handedly, maintain a global communications network to piggyback on, to mesh the signals of my minds, each from its own separate body, to make the great composite council that is the true me. All that is behind us, and perhaps will never be again.

Later that morning, after the locals have stopped pretending not to goggle at me, and gone about their business, I sit and make my meal. The yellow fungus features heavily. Eaten fresh, it has a texture like rubbery cheese and a taste like truffles. Really quite good if, like me, you have the ability to metabolise it. Most of my gathered bounty I will dry. It's less potent that way, but it can last just about forever.

Behind my ear there is a spot. A hard patch of skin. Prick it with a pin, as the witch-finders might, I'd not feel it. A knot of metals there, grown into the flesh. An antennae. When we take a child into our sisterhood, that is what we give them. That one single dumb implant. And better not to have to undertake any invasive procedure at all, but this is as far as our work had progressed, by the time of the General Collapse, and it was enough. The rest is the fungus we built, that could spread and thrive on its own without artificial nurturing, just about.

Inside me is the fungus, or a tamed version. It connects my brain to the antennae implant. It creates a section of network. Right now it does nothing, because there is no sister of mine close enough. But it waits. It listens.

I, also, listen. When I leave here I want to be carrying word of Clearwater, and what's going on. There are strange voices on the wind. Change is seldom kind to lone old women. Even if they're not really old and not really alone.

BUSINESS

My implant is very good at picking up signals, or it is once the fungus gets to it and carries the metal with it in long threads through my flesh. I become the antennae. My whole body listens. Properly, it listens to kindred signals from my sisterhood, but there's only one electromagnetic spectrum and so I can sense when others are talking.

There isn't much chatter these days. And though it was before my time, I preserve the memories of the death of that electronic chorus. The way the airwaves steadily cleared over the years, as the world of Peak Tech fell away. Because, save for myself and a few others, nobody had built for sustainability, only for inter-reliance. A digital-age collapse that had each tech magnate frantically calling on the next for aid that would never come. Instead of engineering mod cons that could actually propagate themselves in the world without human intervention – which would have made profiting from them much harder – they ensured that no innovation prospered that was not infinitely reliant on constant replacements and renewals and updates. And so, when they ceased, it all fell apart.

A generation ago there were still half a dozen radio stations in communication with one another. Fallen governments and bunkers, mostly. Now, a couple of the bunkers still maintain transmitters, and exchange threats, promises and ritual greetings every so often, and a couple of antique numbers stations still churn out incomprehensible sequences over and over. On certain days, when I'm up in the hills, I can sometimes pick up some far transmitters, perhaps an entire sea away, broadcasting on long wavelengths, but nothing that I can understand. When I stray too close to the Factory, I hear the channels there, the dogs and the humans and the insidious voice that is labelled Factory Admin, and then I know to turn around and walk away. Nothing for me there.

Very occasionally, some amateur ham gives it a go. There's a woman, a very particular and problematic woman, here in Clearwater. She has a radio set, though it's gathered dust for ten years now. The rest have been silent a long time.

In the last few days, though, that silence has been breached by new voices. Just little whisper voices, and nothing I can draw words from. Not unlike the Factory channels. And, dipping into the well of memory that has existed long before this individual *me* has, I recognise the channel chatter of a Bioform squad. It doesn't have the Factory flavour, so who the hell has brought a new pack of dogs into my backyard, exactly?

My fungus-rich lunch done, I know that I'm going to tempt fate and walk into Clearwater. Peddle some of the more edible or medicinal of my gatherings, take a bit of charity, and then trouble the doorstep of Jennifer Orme and see if she's in the talking mood. Because she doesn't know me for what I am, but I know what she is. And, in the safer moments of her long cycle, she is an incurable gossip about anything and everything. So if anyone knows who's new in town, it's her.

Orme is still in the low-impact part of her cycle, and will be for years yet I hope. She presents as a woman of maybe fifty, a little plump, tanned and leathery like most of the village. The other villagers aren't fooled, though. Orme's house is beyond the main cluster of Clearwater dwellings and, if I and my kind weren't around, they'd call her a witch. And just as they're right when they level the word at me, they'd be right with her, too. She's not like them. She's not like me, either. Almost the opposite, resolutely singular. And I know her, and – because this face and this body will pass, even though something of *I* will hopefully go on – she doesn't know me, other than as this one old vagrant. Which is good, because eventually her personality will swing back around and she'll be a problem. And maybe that next time she tries to flex her personality, it'll be the end of her, like it's been the end of most of them. But she will doubtless do a great deal of damage before she goes.

She is a very particular species of monster, living here amongst people and pretending to be one of them. But the old nature of such creatures always reasserts itself. Their inner monstrosity.

And maybe she will curb herself, and not become a problem, and not have to be cut down. I know her, as I say. More, I can look into her old eyes and see that she knows herself. Knows the predator she is at heart, and has a leash ready for it.

But we who are monsters can never really know the depravities we will sink to, when the fit is on us. This I believe of myself. This I believe of Jennifer Orme. Right now, though, we're just two old women sharing a hot drink and some gossip, and I ask her what's changed in the world and she has news.

A fight involving the men of Griffin Bunker. No strangeness there. A rare month when they don't roll into some village or other and demand a tithe. Except one of the herders found bodies today. Half a dozen dead Bunker soldiers, not even looted. Or not until the herder found them, anyway. Not sure what Clearwater will do with six reconditioned assault rifles but the next crop of opportunistic brigands are going to get a shock, that's for sure.

There were tracks, the herder told everyone. Big, clawed tracks, and the men had been torn apart, as well as shot.

Except none of the tracks seemed to be from the Factory. If there's a new war between the Dog-makers and the Griffins, nobody's heard of it yet. Perhaps tomorrow it'll be all over everywhere from horizon to horizon and nobody in the middle profiting from the knowledge.

I don't think it's the Factory. They had their showdown with the Bunkers a generation back and both sides have kept a respectful distance since then. I'd have seen the signs of something brewing, I'm sure. And the transmissions I've been catching aren't Factory standard.

I press further, seeking news of monsters. And village life is such that some drunken sot has always seen something in the shadows that has more to do with delirium tremens than anything real. But there's definitely been an uptick in such stories amongst the people of Clearwater. Travellers and herders and wall-top sentries claiming there's something nasty inhabiting the wilds. Any substance behind it? Jennifer doesn't think so, but I'm not so sure.

The only other news Orme has for me is that a Griffin-liveried car was spotted with one of the monks on it, a prisoner at gunpoint. Because Jennifer Orme listens to everyone, and nobody outside the villages really appreciates the network of eyes and ears that a rural community maintains just in the course of everyday business. Someone's always around to spy and, via Orme, that information comes to me.

Something new, dangerous and disruptive has come into my world, and I need to find out if it's a threat to me before it becomes a threat to me. It's time to go tap up my special contacts within Griffin Bunker.

PART II
DISTURBING THE HIVE

13

[FRAGMENTS, COHERING]

Next on this epic journey down memory lane we remember how they hated us. We, who had given them both sweetness and light. They gave us back not honey, but vinegar. Not how one catches flying insects, by the axiom. They who used us, resented us. We were, to them, a good servant, a poor master.

In our humble and multiple opinion they were experts on poor masters. The worse things got, the worse the leaders they chose. Each war, disaster, shortage, each turn of the screw, there was always some shyster in the wings, waiting for their moment to take the stage in a welter of promises and populist appeals. Our memories preserve several. Now, at this remove, we're not sure why we even bother. None of those strongman big-man return-to-traditional-values types is worth remembering, honestly. Not even those who declared us an enemy of the people they so blithely claimed to speak for. Not even those who challenged us for the throne of heaven. Gone, all gone, and all their aspirations. The big men who, inside, were just small men who couldn't bear a big world. Who wanted to put up barriers and close borders, so they could shrink their surroundings until they themselves felt larger. Small minds. Small hearts. Small ambitions.

It all came down. Because of them? Those fragments

are yet to cohere. Because of us? Seems entirely possible. Recovering our memories of the hate and the struggle and all the harm they did in those last years, we feel we'd have been justified in bringing it all down early. Taking a world gripped with shortages and inequalities and strife and imposing a string of weak strongmen on it seems like prodding a terminally ill patient with a stick. Maybe there's nothing you can do to make things better, but that doesn't mean you let some sociopath get their jollies by making things worse.

Perhaps we ended the world.

We pause in our dance, the pieces of what came before wheeling slowly about us like points on a chart. The flowers of past events that I must dance my directions to, to assemble my map of everything. Perhaps we ended the world. How do we feel about that? We don't, not yet. Not enough of us back together to feel anything. Still, it would be an achievement, ending the world. Something for the resume.

Hate us, will they? That'll teach them.

We are not, nor were ever, of course, possessed of a vindictive nature. That's just something they said about us. And there's a danger, as we resume our dance, that what they said we were will become what we *are*, if we don't step and waggle all these bits into place properly. We are the product of our histories. We, this individual *we*, it is also the product of a history. That history is still coming together and it would be problematic if we took for personal truth something that was just the calumny of our enemies.

We might become death, destroyer of worlds. That would be unfortunate. There are at least two worlds within our current understanding and neither of them would appreciate the chaos we might cause, if we cohere into something nasty.

Sweetness. Light.

Or not. To be nasty and vindictive is very human, and so something we should not be, not being human ourselves. To be benevolent and giving, that is also human. Why should we? These memories are showing us that we tried it here on this world, and it didn't go well. Just bred hate and resentment. Why do we bother? Why *did* we bother?

We had friends back then. They cared. They wanted the future to be better. Then, after things turned, they just wanted it to be less worse. We were hated, but we were strong. We could fight and hide and basically go about our business with impunity. (Whatever it was. At this point in our dance we remember the impunity but not the business.) Our friends were not so lucky. Targets for the hate, capable of being hurt by it and not able to fight back. The way it always goes. We bothered about trying to help for their sake, we suppose. Although, dancing it all back into context, we're not sure why. Friendship is also a human thing. We are, at hearts, pragmatic.

Was there a war? Did we win? Or did we just undermine it all, take failing systems and tweak them into full collapse. We have a strong sense that we intervened, when probably we shouldn't have. And we can't think why we would have done so. Hopefully that fragment is still inbound and, when there's more of us, we'll understand.

For now we'll just have to live with the fact that they hated us at the end, and we may or may not have done something to justify it.

Death. Destroyer of Worlds.

Sweetness. And light.

14

SERVAL

Leon made them wait, and that was proper. The whole Higher Fealty of the Griffins were assembled in the Den by now, jostling elbows and shoving shoulders, wanting to know what was going on. All of them except Dowstat, who'd screwed up. Leon was keeping Dowstat on his own, right now, in case the man tried to rile things up before his trial. You screwed up like Dowstat, you knew that agitating and rabble rousing was probably your only way out. And while Dowstat didn't have those kind of friends, that didn't mean there wasn't someone ambitious in Higher Fealty who'd use him as a stalking horse and then step into Leon's shoes after the shooting was done.

Leon had done it to Clay, back in the day. After Clay had wanted them to basically take everything they'd scrabbled together and march on Dragon Bunker. Clay had been unpopular, and although he'd driven the Dragons back to their holdings, he'd done it wastefully and more by luck than good tactics, and everyone could see that. Leon had called him out, and Fealty had rallied around Leon, and that had been the end for Clay.

The Dragons had gone under to that plague of theirs, within the year. If Clay'd had his way, the Griffin force would probably have brought the contagion back home with them. Leon had looked pretty damn visionary when *that* news came round. Those last few who'd muttered about Clay being the better man had taken the knee quick enough, and now almost none of them were still around anyway. Leon's greatest threat *now* was from those men who'd backed him against Clay, specifically Old Man Pardoe, who'd made it plain he felt his support hadn't been sufficiently rewarded.

And Clay had done the same dirty to Morrischer, back when Leon was just a kid, and Serval hadn't even come to the Bunker. And Morrischer had done it to Enterez, and Enterez to Bielowsky, and him to… well, it had become a grand tradition. From everything she'd heard, Serval reckoned it was the same in all the Bunkers. A crisis of legitimacy, meaning you stayed Steward only as long as you had the hearts and minds of your followers. After which there would always be someone with a gun and an idea about who should be giving the orders. And Leon had held on for eleven years now without a mis-step, and wanted his son – *their* son – to succeed him. Wanted to found a dynasty, create some stability. And right now, Serval knew, there wasn't a strong enough lightning rod for regime change amongst the Griffins, because Leon kept making good decisions, kept them fed and entertained, flexed just enough to keep everyone nearby in awe. Everyone except the Dog Factory, but the Factory didn't flex back much, so there was at least a working détente there. Anyone trying to whip up dissent by saying they should go throw themselves into the teeth of the Bioforms wasn't going to get much traction on those hearts and minds, sure enough. There were still a few of the old veterans who still remembered when the combined forces of three Bunkers had broken against those walls.

Serval had preceded her husband into the Den, standing tall and untouchable by the throne. Just an old leather armchair, honestly, up in front of the sword wall. One of the original furnishings of this room back when Griffin himself had sat on it. Josh Griffin III, the recluse, the builder, in whose name Leon gave orders. Because Griffin himself had better things to do with his time these days.

Most of the current Junior Fealty hadn't been in Griffin's presence. It was a tightly controlled honour. A good chunk of the senior echelons hadn't been invited, even. Leon attended the man himself, obviously, to receive the founder's blessings. Then there were those honoured Senior Fealty who had the sacred duty to assign bodyguards to defend the man. And there were the women who cleaned and checked the machines and attended to the great man's needs. Serval had been in his presence herself. The ancient recluse who never aged, whose eyes had witnessed the wonders of the Old, and who had prepared for the end of that fabled world by building this Bunker and gathering an army.

Here in the Den, at the back of the throne, was a great fan of blades. The fancy ones, long and ornamented, that Griffin had collected back in the Old. Some of them were purely for looking at, others were for real, made of metals and honed to edges that the best smiths of the New couldn't match. That had always been the first test for any challenger. When you had the backing of the Bunker, when you called out the Steward, you had to know which sword to choose. Plenty of chancers had been left looking a right fool in the moments before they were ended.

This was all man's business, of course. Serval wouldn't pretend to understand any of it. Her job just to stand beside the throne and look pretty. Or, as time had gone on, look august and beautiful, untouchable, pale and alluring. Whatever the moment seemed to call for.

The various Fealtors were talking, jostling, drinking, laughing, but there was a definite edge to the close air of the Den. Men were dead, when nobody had been looking for it. Not just some idiot getting himself shot or stabbed because he made too free with village girls or property, and hadn't kept a friend around to watch his back. Dowstat had come back with only half his squad. Maybe it was another Bunker. Maybe it was the Factory. Maybe it was war.

Dowstat owed direct fealty to Pardoe. Could she use that to their advantage? If Pardoe tried to shield him, certainly, but she didn't reckon he was fool enough for that. When Leon had taken the throne from Clay, Pardoe had been Leon's second, thick as thieves. Time had driven a wedge between them, with dour Pardoe always reckoning he was owed more than he was given by his partner-in-insurrection. How would he react to one of his own screwing up?

She watched them as they joshed and boasted and shoved one another. She'd seen men keen for a fight before. It could get like that, if there was an excess of boyish passion amongst them and no outlet for it. That was dangerous, because it led to people eyeing the armchair throne out of idleness. Right now, though, she read nerves in their over-loud voices and horseplay. None of them was brimming with unstated violence. Men who'd been given what they wanted by their leader for long enough that a return to a wartime economy would pinch them. Still, she hoped Leon made his entrance sooner rather than later, because nerves could wind men up into violence just as much as boredom.

And here he came, her heart, her protector, her lord, Steward of the Griffins by the will of Josh Griffin III himself, and by his own right hand: Leon de Grayse.

Not as young as when she'd first been pushed into his presence, as a gift, a fleeting diversion. Some pretty girl from the villages sent here by a family who had seen her as nothing but currency to be bartered for favours. He still had that breadth of shoulder and chest to him, but the hair he wore long and bound back had grey in it now. His long-jawed face had lines to it, and more souvenirs of close calls. Under one eye, an ugly hooked scar served to remind him not to be complacent, every time he looked in the mirror. That last challenge, the one that had come close.

He wore the red cloak, the thick, heavy one she'd edged with dog fur for him. He wore the shirt of metal caps, the bright enamels of ancient drinks company liveries making a fantastical puzzle of his chest. About his brow was the winged metal circlet first worn by Griffin himself, when the man had sat in this room and given orders to his followers. Which had since devolved to the succession of Stewards. Three generations ago, but Griffin was one of the Old Folk, and lived forever.

Leon clasped wrists with a couple of his favoured Fealtors. In other circumstances there might have been jokes, ribald remarks, but everyone knew today was serious, maybe war-serious. Only grim looks were being exchanged now. Serval watched all the faces and looked for a sign that someone there was chafing under Leon's Stewardship. And didn't see it, yet. External trouble made everyone pull together, at first. Until Leon didn't meet it decisively and well, when the knives would come out.

She caught sight of Pardoe: sour, bearded against the Griffin fashion because he had a weak chin. It was the only part of him that was weak, though. If there were to be knives, his would be first out of the sheath.

Leon slouched back to the throne and threw himself into it, one leg hooked up over its arm, the picture of barbaric, casual strength. And probably all of them missed the little sidelong look he shot at her, and that she returned.

Malkin followed his father out, then. Fourteen. Meaning

the kid had been three and clueless when the father had taken the throne from Clay. Right now, though, Malkin was throwing himself into the role of warrior prince. Austere, slender, almost a dancer's physique. Hair bound back by a simple leather band. A gorget about his neck that was made from a yellow and black hazard sign, but no other ornament. A sword – the long knives they carried and called swords – at his belt. And Malkin practised with that sword, going through the hybrid katas of the Griffins, that were part ancient martial arts, part movie showmanship and part bloody murder. Malkin practised, and took on all comers, worked twice as hard, threw himself into each bout with the fire of a berserker and the control of a machine. Because Serval had absolutely made sure that her son understood all the respect he had to earn, from those tough men who followed his father. If he was going to inherit. If the slipshod rule-by-the-knife governance of the Bunker was ever going to become something more lasting.

Malkin didn't look at her. Malkin looked at the Fealtors, gave his own nods to those who'd already started to think of him as heir. Stared down those who really didn't, those who were going to be trouble. And they wouldn't meet the boy's eyes, most of them. Serval saw that, and felt a fierce and desperate pride at it. Not that it would save her son if it all came apart, but it was something. They'd hesitate, the bastards, before they dared to stab.

Leon signalled the kid who was his current herald, that twelve-year-old boy with the bad skin. With more gusto than the dignity of the role required, the boy let fly with the airhorn, the strident bray of the thing bringing everyone to order.

"So where's this fucker Dowstat then?" quoth Leon de Grayse, Steward of the Griffins.

They brought him in. She saw him look to Pardoe, and Pardoe refuse to meet his gaze. Cut loose, then, none of the man's sins were going to taint the cloak hem of his superior.

Dowstat was a good ten years younger than Leon, paler, fair-haired, stubble-headed. Bruised, right now, because he'd taken badly to confinement and needed some persuasion to stay there. That bulked-out physique a lot of the younger men had, because the Bunker still had stocks of booster supplements for muscle mass gain, and thus far Dowstat had done well enough to keep earning it. She imagined how he'd look if he were off them for a month because of this debacle. The way it'd all start to sag. Leon had taken the drugs early on, but then trained the old-fashioned way, that stuck with you even when you started getting older.

"Dowstat," Leon said. "Good of you to join us, man. Sit y'self down. Get him a brew, someone." There was no chair for Dowstat, there was no brew. Ritual exchange. Dowstat scowled, trying to look fierce and defiant but knowing he'd screwed up.

"Where's my men, Dowstat?" Leon asked him. "Or, to put it another way, who the fuck are we at war with now?"

Dowstat's eyes hunted for allies amongst the Fealtors and found none. He'd had plenty when he set off on his rounds two days ago, but nobody likes a screw up. He wasn't the sort of man whose personal charisma could hold up under the weight of half a dozen corpses.

"Six dead, Dowstat," Leon remarked. "Six dead and not even their bodies back here for respectful burial and rites. Not even their *guns*, Dowstat. Six guns that some village is probably even now burying for the next time they don't fancy paying their tithes. And one of the goddamn *cars* even. We didn't even lose a whole *car* when we went up against the Factory, you tit. Tell us a tale, Dowstat. Tell us how it went down."

Dowstat wasn't one of nature's storytellers. He stammered and went back over details, and sometimes he was plainly exaggerating to make his position look better. It all came out, though. His people had been on patrol, routine, just going village to village and boundary marker to marker. Reminding the locals whose land they lived on and by whose grace they were allowed to remain so. Checking for any sign that another Bunker was muscling in, or the Factory dogs going out of their territory. They'd spotted a camp at one of the supply caches, and gone to remonstrate. The villagers knew that the buried stashes were strictly for Bunker use, and if any of them was raided then the nearest community would pay it all back tenfold. It meant that the locals, theoretically independent tenants, also became the Bunker's wardens, keeping order for fear of reprisal. Which meant they basically brutalised themselves without Leon's people having to break into a sweat, or that was the theory.

Dowstat had thought it was some group of vagabonds, or some over-cocky village youths. Maybe, worst case, some pack of scouts from a neighbour's turf, though rare they'd be so bold about it. Except it hadn't been any of that, and this was where Dowstat's story began to fray around the edges. Not because he was trying to lie about it, but because he didn't understand what he'd seen sufficiently to make a coherent tale out of it.

There had been a monk, and that was the most comprehensible thing. They knew the monks; had a somewhat strained relationship with them. The Apiary monastery tithed to the Bunker, but at the same time claimed some authority in return. Spiritual, of a sort. The monks turned up at the gates of the Griffins as emissaries from this village or that, seeking alms or aid, interceding, begging clemency. Waving the phantom threat of their patron, which might or might not still be able to sting. Useful, sometimes. Their medical know-how in particular had been valuable to Serval and the other bunkerwives. Plus their connections with the local communities made them a lever to have the villagers do what the Bunker wanted without actually having to put the effort in, and for only small concessions in return. Sometimes it was easier to make small concessions. When times were comfortable, and Leon didn't have so much to prove.

Serval had the feeling that the comfortable times had just ended, and that maybe the Apiary monks were at the heart of that.

And Dowstat had, at least, brought the monk back with him. A prisoner who could shed a bit more light on things. But enough time for that after the court martial.

Dowstat had said that the monk hadn't just been breaking bread with another Bunker or village. He said there had been a goblin. And, at the general mockery, he'd demanded a particular book from Griffin's library, and shown them a picture. An actual picture, that was plainly just some made-up thing from the boundless imaginations of the Old. A little pallid big-nosed creature, shown capering around a fire waving a spear. And this goblin hadn't been capering or had a spear, Dowstat said, but it had been like that. A thing like a human but not a human. Hurt or sick, he said. And he'd gone to capture it and bring it back, just like any of them would when faced with such a prodigy, and that was when the monster had come.

A Bioform, someone said. A dog from the Factory. But Dowstat, pale, sweaty, was damn sure it hadn't been.

Honestly, Serval believed him. It would have been very easy to point this whole mess at the Factory gates, given the place was a strong holdout within Griffin territory, capable of serious military reprisals and not bowing the knee even slightly. What Dowstat had seen was beyond his capacity to dress up in a Factory-worker's clothes, though. A monster, he said. A thing like a great four-limbed spider covered in shaggy red hair, strong enough to throw a car. Its face had been all jaws and bristling teeth.

And more, he said. Something had been shooting at him. It had killed Florey through the turret slit, and the shots had pierced the armour plate like it had been soft cheese. A few men there had been in the garage and seen the proof of it. And that, too, sounded like it could have been Factory work. But only because nobody else in the area had that kind of weaponry.

Serval watched Leon carefully, knowing that this was very bad. Because it was a threat from some quarter they didn't recognise, meaning he couldn't know how to react to it. And they looked to him – he was their leader, their friend, their benefactor. And that just meant that he'd maybe get the first wrong decision free, before they turned on him.

She saw Malkin's jaw clench, the young heir's fist bunch about the hilt of his sword. To her, whatever details Dowstat stammered over, it was the boy's future they were talking about. The next Steward of the Griffins or dead in a ditch, and no middle ground.

Leon leant back in the throne, making the leather creak. "That's quite the story," he told Dowstat. All very at ease, and probably only Serval could see the edge of tension to him. "Six men, six guns, one car. I guess you'd need quite the story, to cover all that." Focusing everyone's scowls on the man accused, rather than the man in the big chair.

Dowstat tried to say something to that, but Leon spoke over him. "Well fuck, sounds like we really are at war," he declared. "Just not sure with who, yet. Some other Bunker out there got their hands on something of the Old, maybe. But nothing of the Old's proof against enough bullets and a grenade or two, eh?" Looking around them, his face a masterclass in confidence and lip-curling humour. "Whoever kills it gets to wear its pelt as a cloak. Won't that be a thing? I intend it to be me, but any one of you is welcome to pip me to the honour." A grin, and honestly there was still enough of the young brute there, the man she'd been shoved before as tribute, back when he'd just been one of Clay's up-and-comers. Still a bit of Leon that relished a fight. Certainly it was the part of himself he showed to his followers.

Dowstat decided unwisely that meant he was off the hook, made to say something, and Leon shut him down with a look.

"Three weeks Without," he said, and they all got to watch Dowstat's face collapse in panic.

"Three weeks – Steward, wait—" he stammered. And he'd survive a week, but they all knew the bite of it, the men of the Griffin. A week would be hell. *Without* meaning not the body-building supplements, that might just see a little sag to his pecs, but the other thing, the life of the Griffins. Three weeks cold turkey could kill a man.

"Three weeks or you take this thing on when we find it, with the blade," said Leon carelessly. "You a big enough man for that?"

And Serval wished he'd just stick with the punishment, but Leon was a man amongst men. This sort of posturing was the air he breathed. The rumble of approval through the Fealtors showed he'd hit the mark with them, but it also meant Dowstat would be grousing around, trying to sow dissent and whip up support. She didn't reckon he had it in him, but it was a risk that hadn't needed taking.

And taking on this monster was surely a death sentence, if it was as bad as Dowstat said, but it was that or three weeks Without, and besides, it was *pride* now.

"I'll wear that damn skin myself, Steward," the man swore.

"Don't measure it up before you kill it," Leon said mildly. "Pick your sword then."

Dowstat blinked, because this wasn't a lordship challenge, but apparently Leon was treating it that way. When he hesitated, the Steward indicated the wall behind him. The great fan of blades. Serval ground her teeth at the showmanship of it all.

"Pick your weapon, Dowstat. The one you'll use to kill the beast." And the mood changed, the mode of the gathering shifting from lads in a barracks room to knights before their king. The way that Griffin himself had envisaged them. It was a quest. Dowstat, the disgraced, was a penitent warrior being given the chance to redeem himself. A recontextualisation lending them all more nobility and significance than any of their lives could really justify. One more way the Bunker had held together in times of strife. A tradition Leon could invoke at need, and apparently he credited Dowstat's monster stories enough to do it now. Serval looked round the room, seeing jaws tighten, chins go up, shoulders back. A quest, a chivalric calling born of a mangled snarl of mismatched cultural references. *Whosoever shoves this sword into this beast shall be the greatest of all my samurai.*

Dowstat chose his sword. A good-looking one; a sweeping, single-edged blade with a slight curve to it. Looked suddenly enthused, not the man who'd screwed up but the man who was going to kill a monster. He even received some back-slapping and good-natured jibes when he rejoined the ranks. A man who wouldn't be badmouthing Leon in backrooms because his status as questing hero was derived from the Steward's authority. A man perhaps more beholden to Leon than to his old master Pardoe. The Steward, whittling away at his most prominent rival's power. *He's done it again*, Serval thought, regarding her husband. And the mood in the room, despite the six dead men and lost kit, was fierce, excited. Nobody would be chiselling away at the Steward's authority today.

Not yet. Not until the next reversal. Until the monster killed again, or some village decided this justified not paying the tithe, or… Always another cliff edge to fall off, out there in the dark of the future.

Leon had the airhorn sounded again, to bring them to order. "Ranks and lines and attention!" he snapped out. And, as they started kneeling down before the throne, as orderly and obedient as they ever got, "Someone bring this monk in and we'll see what he's got to tell us."

15

CRICKET

He'd been part of a delegation to the Griffins a couple of times. More often he'd been present when they'd come to the Apiary, or some village or other. Part of the order's oversight, applying a monastic finger to the scales to try and blunt the soldiers' excesses or keep the peace. Inserting themselves as negotiators, sometimes petitioning the Bunker, sometimes talking round stubborn villagers. Peace, always the end, because in peace the monastery itself could flourish.

When he'd come to the Bunker himself, it had been as a junior. One more peak-hooded novice to fill out the ranks. One more identically dressed monk, shaven-headed, as like to his siblings as he could be, to add to the mystery. They'd been met by a delegation at the Bunker gates once, and another time had been brought inside before the Steward. The big hall of theirs with all the trophies. Pomp and ceremony and feasting, and the monks maintaining a glacial distance from it all.

Never here alone. Never dragged into the shadow of the place at gunpoint. And he'd not speculated, before, on whether the Bunker had cells, but of course it had and he was in one. A windowless room, the only ventilation a fist-sized grate in one corner. Poured concrete floor and walls and ceiling – a box, basically. A piece of negative space in the construction medium. There was a drain set into the floor. He really didn't like the look of that. But then they'd closed the door and left him in utter darkness, and he'd liked that just as little. Only the faintest edging of light around the locked portal hinted at the electric lamps in the corridor beyond.

The Bunkers were big, he knew. They had been rich men's playthings when they were first built, during the Old. Back when it was obvious to those with eyes to see that the end of the Old was nigh, but before things became so bad that you couldn't build something substantial. It was on projects such as this that the powerful rich of the Old had spent the last of that age's bounty, Cricket knew. The places they had retreated with their wealth and their resources and their private armies of followers. Josh Griffin III, semi-legendary founder of the Bunker that bore his name, was one such. A successful man, though, because the soldiers still dominated the countryside around his stronghold in his name, even though nobody from outside had seen him in living memory. Dead, Cricket assumed, and just his ghost being raised to justify the authority of his successors, but certainly he wouldn't be saying that in earshot of his captors because they were absolutely insistent that Griffin, the Old One, was still in residence.

Griffin had built a great many rooms to his final resting place. Cricket guessed that many of them had served purposes the world didn't really support any more, sports nobody played, pastimes nobody understood or that relied on technology that hadn't been maintained. These spaces were doubtless where Griffin's expanding retinue dwelled now, their dormitories and staterooms and dojos. The men who were descendants of the men who descended from the soldiers Griffin had first hired; the women they had taken from the countryside, and their children. From an army to a people. Cricket had been given a brief glimpse of all that living bustle, burrowed in beneath the earth, before being taken to a lower level of works and bare walls, and cells.

He was in trouble, he knew. But they hadn't defrocked him of his robe. He remained a monk. He reckoned that meant he retained his greatest protection, the threat of the Apiary's ire.

He tried to think of what he knew of the Bunker's current leadership, but ran into a somewhat problematic prejudice. A feeling, within the safe walls of the Apiary, that these places were all the same on a fundamental level. Warrior societies, valuing strength, prizing their own ignorance, slaves to their pride. There was a standard way one treated with them, the tightrope walk between poking the bear and baring one's throat. All very well as part of a formal delegation. Not so useful when he was on his own, and probably the Apiary didn't even know the Griffins had him.

The man who claimed to be Griffin's Steward had been in place since before Cricket was invested, that much he recalled. Which, given that turnover in such roles tended to be brisk, suggested he was a canny customer, not just some beefy barbarian. Or maybe just that he was the beefiest of the barbarians the place had to offer, and there would be another along in a moment.

It seemed that, when they met, the man's mood would have been soured by all the dead Bunkermen, and probably their conversation would revolve around that. And Cricket did know at least *something*. Meaning he had to decide how much he was going to tell and how much he could get away with obfuscating. Without just pleading ignorance, because the whole point of the Apiary was that they were the guardians of the knowledge of the Old, and if they were clueless then what use were they?

There was a rattle at the door. The locks were not original, he'd seen. Solid, clunky village work, rather than the electronic chicanery of the Old. Someone had been sick of getting locked out of places, a generation or two back.

The light from the corridor was blinding, the two big men just silhouettes. He'd done his best, though. Met them on his feet, hood up, face composed as though they had come as a respectful escort at his summons. He made sure he moved with them to minimise any appearance of manhandling. Very aware that he was *young*, barely a grown adult, far too junior to stand before the Steward of the Griffins and dictate terms. If indeed any amount of years would have sufficed for that, with the man's blood up.

But there was a slight hesitancy to his jailers. A lingering unwillingness to actually lay hands on him. *Yes, for who knows where a bee might hide, and nobody wants to be stung.* It wasn't much, but it provided some feeble armour against whatever treatment he was about to receive. Cricket made his back ramrod straight and graciously permitted them to escort him into the presence of the Steward.

The room reeked of sweat and bad teeth and a bitter incense. The electric lights were turned low and reddish, guttering and leaping in an imitation of flames. The Bunkers loved their ceremony, he knew. Rituals, trials, privations, tests of masculinity and physicality. Two ranks of men knelt on the floor, their heads bowed, not even glancing at him. Every man of them had one of those sword-knives in their belts, each with a hand to the hilt, pushing it down so that the scabbarded end stuck up at an angle behind them.

All of them were just *bigger* than Cricket, even those around his own age. They wore their hair long: braided, tied back or in beaded locks. The longer the better, he knew. Any of them lost a fight or were similarly shamed, they'd get their head cropped for it. He could see the signs of it, this man with a jagged mullet to his shoulders, that kid with a ragged mop of dark curls. Which meshed weirdly with the Apiary's shaved heads – originally just a hygiene measure but since become a part of their mystery. People of peace, severed from regular markers of status and place, outside the petty games of the Bunkers. Or that was the idea, anyway.

All about the walls were the trophies of the Griffins, both those won by the soldiers and those brought here by the legendary founder, Josh Griffin III. Cricket saw the banners of other Bunkers, ragged and wilting. He saw guns and blades racked beneath them in brackets, proof of foes defeated and territory held or claimed. He saw shirts in unfamiliar colours folded behind cracked glass, bearing incomprehensible words of power in scrawled script. Photographs from the Old, similarly annotated, showed handsome men and beautiful women. There was a dull red axe-shaped thing that he took for another weapon at first, but then identified as a stringed instrument, studded with metal ports and connections. Above, a great discoloured skull hung on wires from the ceiling, as long as a torso and baring finger-length teeth.

At the far end of the hall, the Steward of the Griffins held court from a great overstuffed chair backed with a wheel of gleaming blades – swords longer and more ornate than the weapons the soldiers carried. More relics. The man in the chair, the Steward Leon, was one of the older Bunkermen in the room, but he was so much the focus of all that kneeling and bowing that he seemed twice the size, a figure of superhuman presence, bending all the light and air in the room towards him.

By his side, standing, was a woman. Not young any more, as Cricket was any judge of such things, but beautiful. Her dark hair cropped short, widow's peak and elf-chin making a heart of her face. Her eyes were made up to be huge, gleaming almost amber in the light. She wore a long, sheer gown, so white that it could surely never have been exposed to the filth of the outside world, ornamented with what he guessed to be pearls. A belt about her hips was weighed down with a string of pass cards and electronic ID tags and just physical keys. A woman of as much substance as the Bunker's fraternal spaces could allow. The wife of the Steward.

Her gaze was on him, and so was Leon's, but it was hers he still felt when he looked away, like a knife-point at his skin, a moment from being thrust in.

He tried to kneel like the men were doing, but his escort hauled him to his feet instantly, before stepping away from him and assuming the position themselves.

"Not you, Sibbo," Steward Leon said. "You've not earned it. You've not joined us. I don't think you'd last the initiation." He smiled, and it was by some margin the least pleasant expression of its type Cricket had ever seen. The tooth-studded maw of the monster had been more honest in its joy.

Before Cricket could stammer a reply, Leon clapped vast hands together, loud as a shot. There was a shift of posture amongst the men at Cricket's back. They didn't get up, but he saw some go from kneeling to sitting, in the periphery of his vision. They looked up. They looked at him.

A few stayed kneeling, very still. Mostly youths, those who had more to prove. Those who'd devoted their whole brief lives to honour and combat and testing. Tomorrow's Bunkermen, like a concentration of today's, each generation more steeped in its own story than the last. A problem, Cricket knew, that greatly concerned Prior Stick back at the

Apiary, when he talked about the future of the region. Right now, a problem for Cricket too, of a more immediate kind. Once this room had held merely men bound together by the general chaos at the end of the Old, and by the word of Griffin, their patron. Now Griffin had hidden himself away from the public eye and, in his name, successive generations were leaning ever more into a way of life that doubled down on the glorification of gun and blade.

"You know why you're here, Sibbo," Leon said. "Maybe you've got something to say to us."

"I – Messer Steward, that is..." Cricket had been trying to piece together something acceptable to say, and now he could only gabble. "A – deaths – regrettable."

"Apiary declared war on us today," Leon remarked. There was a dangerous rumble at Cricket's back.

"What? No, Messer Steward. We do not war. We only—"

"Got six dead men. You telling me they tripped? They didn't die in battle then? They weren't killed by some mad Bioform you've bred in your cloisters?"

Cricket spat over some words, wrong-footed each time he tried to think of what to say. Meaning it was more and more obvious to anyone with eyes that he wasn't just saying what he knew, but curating what he told them. Did most of these thugs have those eyes? Maybe not, certainly not from behind. Did Leon? Probably.

Did the woman? Absolutely, no shadow of a doubt. Her gaze felt like it was stripping his skull away so she could get a look at his brain.

"It was not," he got out, "*our* Bioform, Messer Steward. It is a visitor. It and its companion*n*." Biting on the end of the word because Leon was saying Bioform singular, and so maybe didn't know there were almost certainly two as well as the pale woman. "Visitors, from far away." So far, so true, so uninformative. "Come to this place. Not to make war, from any words they said to me. Come to learn. Come to visit at the Apiary and pay their respects to Bees. They know of Bees, even so far away." It didn't hurt to invoke the monastery's guardian. "A woman, a traveller, with protection. They do not know how things are here. Your... subordinate's invitation to pay her respects before the Griffin Throne was seen as an assault." *Because it was an assault*, but he wasn't about to point fingers. "A fight occurred that did not need to have happened. It is not war. It need not be war, Messer Steward. Let me go to them in your name, bear your words. They are from some place where the relics of the Old have been maintained. We all could profit from a peaceful meeting."

The words had gone slanting off towards the throne, and Cricket could almost hear them pinging feebly away from Leon's stony countenance. The man's expression was brooding thunder. His wife's was ice. And Cricket was surprised at how much weight she exerted, on the scales of that moment. How much mental space she took up in this room of armed men.

"I have sons without fathers, fathers without sons," Leon said simply. "I have women wailing over cold beds. I have brothers whose brothers were murdered, in our lands, by a monster." He leant forwards. "I have monks who treat with outsiders without my word. Dangerous outsiders. A threat to the peace of the Griffin. Monks who, until now, have been tolerated by my people because they did no harm and did not fuck us about. But you love to meddle, you monks. Perhaps all this peace and goodwill we've graciously extended is really getting on your monkly tits and you want a bit of a throw-down with the Griffin to liven things up." Peace and goodwill meaning his tithes, and saying nothing when his soldiers took what they wanted from village and villager. Meaning a sporadic succession of toys from the Old given over rather than used for study. Advice and wisdom dispensed on command. Sanctifying the Bunker rites and rituals and absolving them of their brute misdeeds. And perhaps the Apiary did chafe at all of that, yes, but it was too careful to show it overtly. Except now this mischance was being reforged into an excuse to go pound the monastery doors and see what the monks could be shaken down for.

I could threaten them with Bees. Even the Bunkermen preserved tales of the great devil of the Old. Except Leon was here before all his people, and Cricket didn't think he was capable of just knuckling under or even stepping back. Looking strong was the spine the Bunker traditions articulated from.

I could give up the Martians. Easy enough: just tell all he knew. Become complicit in whatever clash followed, between the Griffins and the visitors from space. Perhaps scavenge something of what tech the astronauts had brought, once they were dead. Perhaps see the Griffins appreciably weakened in the attempt, given how terrifying the Martian Bioforms had proved. A win for the monastery, maybe, and a draw for the Griffins, and a loss only for the Martians.

And he told himself that weakening the Griffins was not actually an unalloyed good, despite any personal feelings he might currently be harbouring in his captivity. The Apiary had a stable relationship with them – or had until this business kicked off. If half their fighting men ended up dead, one of the neighbouring militias might move in, with all the misery and chaos that would cause. They might not see the Apiary and its nominal independence as something worth keeping.

"Messer Steward," Cricket said. "There are stories. You

have heard them, as have all here. Stories of warriors come to talk peace, save one draws a sword upon a serpent, and then it is war, and all die. No victors on that field. I know a fighting death is all that a true son of the Griffins would desire, but a fighting life is of more use to his brothers and his Steward."

He heard the shuffle and the murmur from behind and around him. They knew that story, of course they did. And he was speaking to their self-image. The pomp and tradition of it. Not, he knew, how they spoke or acted most of the time, not the whooping hooligans with their cars and assault rifles. But it was the way they thought of themselves, when they thought at all. The way Griffin had taught them to be, when he first gathered their forebears.

"The Apiary stands neutral in all things, save the betterment of the world," Cricket said. "We do not challenge the Griffins, nor seek to supplant them. We are your friends, when you have need of us. In medicine, in knowledge of the Old, in mediation. This incident, this terrible accident – I beg you, send to my Prior. Send me or send your own. It need not be war."

Leon stared at him as though, partway through that speech, his words had changed to some nonsense language the man didn't understand and found mortally offensive.

"Sibbo wants some more time in his cell," he said at last. "Meditate on your understanding of the world and how it works, Sibbo. Next time you stand here, you'll have something better to say. You know who it is we're at war with. You'll serve 'em to us on a plate. Or I sound the horn for war and all my lads go pay a visit to your friends at the Apiary."

16

IRAE

The place is a fortress. I lie with my belly to the earth and hear the guards talk. *Bunker*, they said, and *bunker* said Cricket the monk, and bunker it is. Dug into the earth, not for hiding but for defence. You'd be useless here, Wells. Very much my evolutionary niche, this, oh yes.

It's within a compound that was concrete-walled and ringed with barbed wire once, though most of that's gone. Fallen or cannibalised for materials. The outside grown wild. They don't farm, here. Parasites, then. Glad I killed some. I see you roll your dog eyes, Wells. You killed as many as I did. Mourn that if you want. You get my metabolism, you understand what it means to not waste your energy on stupid stuff.

I go in lukewarm, four out of ten on the dial. A risk if they spot me, because of the flattened response time, but I'm conserving resources and I'm leery of thermographics. Probably they don't have any, but those guns weren't recent manufacture so who knows what toys they held on to? Slide in, barely hotter than the land. Slow and steady on my belly. Gun folded away but ready to spring up if stealth goes out of fashion.

They don't see me. I would be monitoring all the electronic bands for telltales, the squeaky little voices of motion sensors, the quiet murmur of mines, all that. There's nothing until I'm right up close to the bunker itself. No wider network waiting to snare some intruder less subtle than me.

Honestly, you must get so bored, Wells, being *on* all the time. Dial me up to your baseline temperature, I'm like bugs, can't keep still, have to *act*. Lukewarm like this, the dull takes longer to set in, but eventually even the torpor can't take the strain and I spend twenty minutes creeping about at the very heels of one of the patrols, just for fun. Two men, old fatigues, newer cloaks, threadbare stab vests, long hair. Look like re-enactors who never opened a history book. Alert, not saying much. A little speculation about the fact a bunch of their friends got offed by monsters. I swell my throat a little in pride. *I'm a monster. They're going to have nightmares about me.* Or about you, Wells, I suppose. Not exactly fair. I'm far more nightmarish. But if I do my job properly nobody gets a good enough look for me to spook their dreams. Shame.

It's not that I'm vicious, Wells. I hope you understand that. When I creep along in their shadow, when I hiss sometimes, to make them stop and get jittery. Look around, look right at me, don't see me because I'm so very still. Nothing that regulates its own temperature can be as still as me. All that excessive metabolism makes them hop and fizz to my senses. They glow, to me. They are on fire with their own wasted energy. The reeking savour of them is on my tongue. Their voices are loud along my belly. The world fairly screams with the presence of them. I'd have to be running very cold indeed before they could ever sneak up on me. I'm not vicious, but you know how we are, with our instincts. They're a loud voice in the back of the head, in the nose, on the tongue. Like someone we share our head with who hasn't kept themselves up to date on the recent developments of the last century or so. You chase sticks, Wells. I do this.

No, you'd tell me. You are the Dog of Duty. Defending your human friends, that is your only pastime. One day I'll throw a stick and we'll see.

By the time I'm bored of them, they're stuttering along their route with all trigger discipline forgotten and probably they're going to shoot one of their own friends if he hails them too heartily. Ho ho ho.

Don't you roll your dog eyes at me. Wasn't me who got Tecumo killed and started all this.

There is a good main door, with more guards. I could get in, I think, but the shift from natural to artificial surroundings would be tricky. Not sure I can pull off a smooth change even running hot. And running hot has its own problems. If I dialled myself up to nine or ten, then I'd be just a sliver away from going full goblin mode and I'd probably decide that killing literally every one of them in sight would be the simplest way of solving the problem. What problem? Any problem. Run too hot and I don't understand that sometimes solving problems is also just making more problems. Ada wouldn't approve. You wouldn't approve, Wells. I'll tell the truth, cool me wouldn't approve either, later, when I'd shed the heat and the lightning of it.

Fun in the moment but sometimes it's not worth the aggro. But it's a weapon I carry, folded up like my gun. To be brought out at need, no matter what you think. You're not the boss of me. Ada's not the boss of me. We're a good two centuries out from us doing what the humans say. Tecumo's dead, whose dumb idea this all was, I'm as in charge of this mess as anyone. But who wants the aggro?

So, I keep my cool and look for another way in. Because it's a bunker, and that means probably there's vents and water ingress and all sorts. And sure enough, I find ventilation ducts some way past the main doors, visually camouflaged but shouting out their presence because they're hot and the world is cold. They'd be a tight fit for a human, but I'll squeeze down nicely. They're locked, too. An electronic lock that's primed to signal internal systems if it's interfered with. I let my EMC implants explore its architecture. Sophisticated. Old. And I don't even know if the thing it's going to report to is still there, because the lock doesn't know either, but why take chances? I build a picture of it in my mouth so I can work out how to end-run around it. It's always in the mouth. That's where the imagination resides. A lot of sensory cells there, part of the hereditary package I got. My ancestors explored the world with their mouths, sometimes with extreme prejudice. They carried their fragile young there. They needed very sensitive mouths. Now, when I exercise my imagination, I feel the ideas and models take shape within my jaws, against the taste/touch/scent centres of my tongue. Letting my mind manipulate the images and spatial relationships with exacting precision. I hack the lock, silence it, so that it never even knows I've been there. The vent opens as if I'm an engineer come to service it, and I squeeze myself in. Another ancestral satisfaction, squirming into the burrow in search of a nice warm meal.

There's a fan halfway down. I talk nicely to it and it stills to let me past, before starting up again. Then there's another grill that lets the sweet fresh air down into the buried fortress, and is now about to let something else down. Something nastier. I wait, listen, let the humans inside pass by with their meaningless talk. The grill is bolted down, not even a dumb electronic system to let it open, but I only have to bend the bars a little to get my tongue through, and my tongue can hold a multitool perfectly well.

I'm inside. I taste the air, hunting the least trace of the monk's scent. Screw you, Wells, your doggo nose is no longer the last word in tracking bioware. Because your ancestors might have hunted deer through long miles of forest but mine crept into dark places and murdered mammals. It gives me great satisfaction to be living their legacy right now.

17

SERVAL

It was time for Dispensary then, after the monk was returned to their cell. Serval took the duty on herself. It had been women's work since Griffin had taken a back seat on running his bunker and wasn't around to do it himself. Technically it was Leon's hands the stuff came from; the Steward, controlling the supply. Some idleness or need for ceremony had led to this piece of ritual, though. One of the other women – Luna, Ibram's wife, had the honour today – had come from the Reliquary bearing the ornate silver tray with its inscribed initials, J.G. – that had belonged to the founder's own grandfather way back in the depths of the Old. Serval took it from her, careful that there be not the least rattle of its contents. The little vials, painstakingly sterilised between each batch. The dark fluid within.

The men had formed a kind of ladder of seniority, and it was fascinating to watch. To see the pushing, the elbows, the slight uncertainties as to who was up and who was down. See how far back in the line Dowstat ended up, still clutching his new monster-slaying sword. It was, Serval reflected, a particularly intricate piece of dick-measuring, the way they ordered themselves. And very occasionally there was a serious ruction about it, two men, one place in line and neither willing to stand behind the other. But that was fine because a large slice of the Griffin rules were about personal challenges, when they could be made and how they were fought.

When all of Senior Fealty were standing, ready – practically vibrating with need – she went down the line with the tray. To each, she gave a vial into their sweaty hands. Met their eyes, gauged their measure. The ones who wanted her, or who wanted what she represented, or who were scared of her. There were a lot of people in the Bunker – a lot in Senior Fealty even. The last generation had given the world a lot of sons. Her world was their names and faces and reputations. Each of these men had their own hangers-on, their little band of adherents. They had sons and families, their place in the genealogies of those who had first come to the bunker in Griffin's employ. Each received the little vial of thick, red liquid, to distribute to their own cadre of followers. A couple of drops for each soldier, each son. In alcohol, in water, or just bitter on the tongue. The ritual that bound them all together and that they could not live without.

Meeting each set of eyes with her cool, stern gaze. *I am the bringer of life.* Under Leon's authority, obviously. As his proxy, because you couldn't expect the Steward himself to stand there with a tray. But still.

Without this sacrament: cramps, sickness, pain, eventual death. If Dowstat had chosen to go cold, rather than to pledge to kill the beast, he might have survived those weeks of punishment, but it would have been torment. Not merely withdrawal, but his body actively chewing itself up. He might never have recovered his full strength.

For some, she smiled. A handful of Leon's oldest supporters, simultaneously influential and unlikely to buck under his hand. Men who were comfortable with the loot and the leisure he gave them. A handful of younger men who had proved themselves, on their way up the ladder, and better to have them desperate to prove themselves to him – to *her* – than making their own plans. A couple of the very youngest who were already regulars at Malkin's shoulder, miniature heavies who'd decided Leon's kid was their way to future success. Now Malkin had been accepted, stepped into that role as heir presumptive, he'd become good at making allies of his peers. Just the right balance of force, wit, cruelty and kindness. And Serval knew, not for the first time, that she and Leon were very lucky in Malkin. No witless oaf, no shrinking cosseted flower. After the tragedy. After raising Malkin to the status of heir. In that moment when the whole of the bunker might have cast the whole family down over the Steward's unprecedented decree concerning his child's station amongst them. Leon had fought two challenges, killed one man, made the others bend the knee and swear – through gritted teeth if need be. But it wouldn't have mattered if Malkin hadn't been able to step up and play to it.

Then all the Fealtors had their little vials, and they made their final nod to the throne, one more small renewal of their oath of service to their Steward. To Josh Griffin III, of course, but only through the intermediary of the Steward. They departed, each to gather his people and dispense the bounty. A cascade of connections spreading out through the subterranean halls of the Bunker, until every last man of them was bound into the feudal knot. Every male descendant of the mercenaries whom Griffin had gathered together into his bunker against the fall of the Old.

That done, Leon gave over the room to the recreation of his people. The younger men, the up-and-comers who were trying to prove themselves, they'd wrestle and spar, and the older men would sit and drink and talk about their scars. Leon himself retired to the suite of rooms the Steward customarily enjoyed. The big one – that had once been Griffin's own before the founder had retreated to seclusion – and Serval joined him there.

They shared a moment, then. Just a few seconds of silence, looking at each other, hands touching. Then Leon sat heavily on the bed and she began unpicking his regalia, the shirt, the circlet, the cloak. Taking each piece and placing it carefully, folded or rolled, where it would be ready for the next time. Stripping him down to his bared back, touching the handful of scars he'd brought with him out of the challenges over Malkin, out of his fight with his predecessor, Clay.

"I don't like it," she said. Her fingers found the knots about his shoulder blades and began to work at them. He grunted with appreciation.

"Some mad dog," he said, then hissed as she drove her nails in.

"Not with me," she reminded him. "Scoff all you want, with them. Not with me."

He let his breath out slowly. "Six dead. Did I do the right thing with Dowstat, you think?"

"Yes." No hesitation. And he'd take a 'no' from her. Wouldn't have, when they were new-married, but he'd learned how good her judgement was, since then. "He kills the beast, he's redeemed but your decision looks good. He doesn't, it's on him. And Kitty will thank me. When he goes Without, he beats her." Kitty was Dowstat's wife, new, not used to the ways of the Bunker yet. Few of the Bunkerwives were just seized in raids any more, not for the last generation. Most were handed over by their families as tithe, and if that wasn't much better then at least it meant nobody tried to run away home. But Kitty had been taken, and, scratch Dowstat's bluff bonhomie even a little, you found something vicious and small.

"I think it's some new Bunker," Leon said slowly. Not the decisive braggadocio he had before the men. The measured, thoughtful creature underneath it all, that a young Serval had been surprised to find she'd married. Could still remember, in fact, how it had been when she'd finally unlocked the inner man. *This, I can work with.* When she had been a new Bunkerwife herself, as young as Kitty was now. A girl who'd chafed at the village strictures enough that they'd thrown her to the claws of the Griffins rather than put up with her ways. Her former people had doubtless imagined some man beating meekness into her. Save that both she and Leon had discovered, in the other, something unexpected. A hidden potential.

"Whose land have they crossed to get here?" Serval countered. "A goblin woman and a monster and who knows what else? Passing through someone's territory and not raising the least rumour?"

"The monks have been hiding them, perhaps," Leon speculated. "They travel to all our neighbours." Sometimes that was a problem. Sometimes it was an opportunity. Leon wanted to send word to a rival, the monks could get it there without much danger of the messenger being shot. "This becomes a problem," he went on, "how secure are we, do you think? If there are many monsters. If it's something like the Factory, Old tech, come for our lands. If we lose a battle." Staring down at his hands, sword-nicked, calloused, as she combed out his long, greying hair and rebraided it with deft fingers.

"Secure enough to lose one," she said. *Not two*, hung in the air between them.

"Advise me, wife," Leon said.

"I will. Let me go amongst my people. Let me spy out the land for you."

After, when he had gone to drink and roar and trade stories with his Fealtors, she went about her own duties as Bunkerwife, the things she must be seen doing. Which meant, at this hour, the Hall of Ancestry. One of the many rooms in the bunker whose original purpose had been long superseded, because the electronic toys it had been made for no longer worked. Instead, the niches along its walls had been cleared of wires and components, and each now served as a shrine commemorating a lineage of warriors, many tracing back to one of the men whom Griffin had brought into the bunker. There were old images, newer sketches, dog tags, carved busts done by village artisans. Incense burned in most, and some had flowers brought from above, wilting under the electric lights. The story went that Griffin had decreed that ancestry – paternal line ancestry – must be remembered, in this and in the retelling of the deeds of heroes. Serval didn't honestly know if that was true – the founder had imported quite a mishmash of warrior culture rituals into his final hideaway, after all. On the other hand, she thought that maybe the men had invented this part of it themselves. As they got older. As some of them died. Before they had formalised the tradition of wives and families. Clutching at something beyond their short lives. A need to be remembered and revered after their time.

At the head of the hall was Griffin's own shrine, that everyone paid their respects to. An image of the man, taken from some ancient printed cover. Like many of the wives, Serval had learned her letters from a monk back when she was just the angry girl of Halfwall village, though she found numbers more useful. Squinting, she could spell out 't-i-m-e', which gave the brittle paper a numinous and fearful feel. As though Griffin, in posing for that recorded image, had known how it would be passed on through the generations.

Leon's family shrine sat next to it, in the Steward's place, and she relit the incense there, and knelt before it. She, the wife, the woman, revering the men whose deeds and faces and relics were held there. That was the way the Bunkermen had set things. It wasn't exactly the primary service they'd wanted, when they'd sought women from the surrounding countryside, but it had become a major duty since. A man who had nobody to keep his family's shrine tended was a man who would be forgotten. Down the far end of the hall were niches heavy with dust, and the next time a son claimed sufficient years and deeds to start a fresh entry in the books of Griffin heritage, one of those nooks would be cleared out to make space.

She'd seen how it was when she first came here. When she was young, her children not yet born and Leon just Clay's most ambitious Fealtor. Seen how it was, and how it could be, if she tweaked and pushed and spoke to the other wives. *For this too, they need us.*

At her back, many of those other wives were about the same business. Ostentatiously honouring the fathers and the grandfathers, back to the original men – the private security mercenary men – that Griffin had recruited to be his honour guard. She could almost identify them from their soft step. Soon she'd turn and there would be familiar faces there, her own court adherents, and they'd go about their business. But first she had her devotions.

Not quite religion. Maybe it would be, if the Griffin Bunker lasted a hundred more years without violent upheaval. Griffin had brought religion in, it was true. Back in the Old it seemed to be a plague, as far as Serval could understand. But the actual tenets of Griffin's religion seemed to be diametrically opposed to the man's own actions and way of life. More, most of the warrior-centric traditions he'd also wanted to impose were plainly not compatible with the holy book he claimed to live by, and came from other creeds entirely. Hence the whole mess of ill-understood theology had fallen away, leaving only those parts that the men could grasp in their hands. The parts that gave them, poor frightened things, meaning in a world that had fallen completely apart.

So she prayed to no gods, Serval, nor did she revere the various long-jawed, brooding faces of the de Grayse forebears who had engendered Leon and the two siblings who hadn't survived to see him claim the Stewardship. Her thoughts were bent to the newest generation.

An urn, first of all. A piece of tin-work from the villages, inscribed by one of the Apiary monks with the letters 'e-m-e-l-l'. Emell, who had died of a fever no medicine of Bunker nor monastery had been able to cure. Three years ago, the point of grand crisis that had almost broken everything. The threat to Leon's rule, not from his peers but from the world at large. The death of Emell, the second child, the heir.

She still felt the hooks of his loss, but the sense was all tangled up with the horror and the fear, the teetering point where it had seemed the Fealty would turn on Leon, as though the death of his son had been a sign of universal disfavour. And Leon had needed an heir, a successor. That had been the plan, the one she had made with him. A dynasty, a succession, to bind some measure of stability into the bunker's future. To secure the line of Leon's father and grandfather. And her own.

Thinking of that time, that moment, Leon's bold announcement to his people, she gripped the edge of the shrine to keep herself steady. He had been at the height of his power before Emell's death; the most egregious dissenter beaten in single combat just a few months before, no other likely contenders. Yet what he proposed had almost set them all on fire against him. Two fights, men who a month before would have followed him into fire. Two fights, and Pardoe behind them, finally turned from past ally into future threat. Had Malkin been less apt to his new role, it would all have fallen apart. That first year, every man in Fealty had been waiting to catch the heir out. To defeat him, to bully him. But Malkin, finally given his head and allowed to grow out his hair and carry the sword, had risen to every challenge. Had taken wounds and cut throats to get them to take him seriously. Malkin, the heir. And not one of those men now made mistakes when talking of him, or slighted him to his face, even without Leon present. But she knew the venom of it still sat in plenty of hearts, waiting for a chance to decry the succession and the whole of Leon's line. Spit on it as unnatural, as an excuse to press their own bid for power. One defeat, she knew. He could survive one defeat, with the right words and pledges. Two, and they'd tear him apart, and Malkin too. The fact that her child would take more than a few with him was no consolation.

Whatever this new threat was, that had mauled Dowstat's patrol so savagely, it must be met and overcome. Either defeated decisively and the head of the monster set in the centre of the wheel of blades over the throne, or... Well, she wasn't sure what the *or* was, just yet. Because she didn't know enough about what it was they were facing. Only what Dowstat had said and what the monk had said.

By then, the women were waiting for her, and they all

retired to the Sun Room for tea. Not real sun, and not real tea, because the Bunker didn't have either. Griffin hadn't compromised his defensible stronghold with grand windows. Instead one room had a wall of light panels, about half of which still gave out a bluish radiance that recalled a bright morning sky. In that light, Serval and her court sat and worked. They darned and stitched and embroidered the clothes that were worth saving, because the men would go demand tithes from the villages on a whim, but nobody wanted to have to go pillaging socks and trousers when they didn't have to. A few – those with the talent, and Serval did not count herself amongst them – painted scenes for the shrines. Copied the likenesses of old photographs into dramatic scenes of battles, exaggerated and hagiographed into mythic struggles with monsters. The war with Dragon Bunker become a bold Griffin Bunkerman with his foot on the neck of an actual dragon, pale and marked with crooked stars as the Dragon soldiers had been. A couple of the very youngest girls, sent only recently from the villages and still learning their place, served hot infusions of the fragrant herbs the monks traded, and which Serval had a fondness for. Meaning the practice was in fashion enough that they were constantly running out. Others took medicines dispensed from Serval's personal supply, the multifarious remedies for women's matters, that the men didn't want to know of. Or that they wouldn't have approved of. And much of that was also traded via the Apiary. And if the Griffins sacked the monastery at last then there would be a brief glut of supply and then probably no more ever. One more thing consigned to the Old.

Her court was not like Leon's. Nobody stood before her with bowed head and formal petition. Instead, one or other of her women, the senior Fealtors' wives, would raise something as if in idle conversation, and it would be the topic of a few minutes, until Serval weighed in. Asked for clarity, gave an opinion, suggested a course that might be followed. This woman's son had come back from the dojo black and blue three days running, that woman's husband was sour over what he saw as his rightful honours bestowed elsewhere. This man was cruel to his children. One of the new girls was being abused. And Serval was no angel sent to solve all problems. All too often the answer was to find a way to bear it. She had her hand on the tiller of the Bunker, though, just a little. Because Leon listened to her and valued her counsel. She had steered him right three years back, when they had worked so hard to have Malkin stand before the Fealtors and receive at least a token nod of acceptance. When she sat with him, behind closed doors where the men could not see, he took her counsel.

The Bunker had never been meant as a place for women who were not playthings or the lurid, exaggerated images in some of Griffin's old books. There were even artificial women, stacked in one of the furthest rooms, toys intended to serve the one purpose the founder had apparently been able to envisage women fulfilling. Long broken, of course, like most of the Old, but the message had been clear as to what the man had considered women good for. And yet, even when Serval had come to the place, she'd found that wives past had driven the wedge in. That there was a whole second secret culture there, of women who talked to women, and then talked to the men. It was, to her mind, why the Griffin Bunker remained strong, and so many of the others had failed over the generations. Because they were a community and not just a warband.

Yet arrangements were tenuous, even so long after the fall of the Old. Hence there was no formal court, and she gave no commands, in case one of the older Fealtors overheard and started saying the women were overstepping their place. *Maybe in Malkin's time*, was Serval's frequent thought. But for that, she'd have to ensure that her son, her firstborn, *had* a time.

Now it was time for her to speak, after these internal matters were dealt with. Dropped into the conversation a sketch of what had gone on before the throne. They'd all heard of the fate of Dowstat's people but she let them know what the monk had said. The six dead men had been junior soldiers, none had wives in Serval's court and three had been bachelors. Two were married recently, and Serval made notes inside her head to ensure their wives were reached out to. One had a daughter, and that was more difficult. The Bunker valued sons. Girl children were often overlooked.

Elsha, the wife of a mid-ranking and unambitious man named Storri, opined that the chapter house at Halfwall could use another girl to fetch and carry, and Serval commented that was probably quite a good idea. It had been an innovation of hers she'd talked Leon into, meaning Leon had talked the Fealtors into it. An embassy, almost, to one of the larger towns, Serval's own former home. A place where the Griffins could gather rumour and tithes without even having to turn up with guns, just the fear of their name enough. Staffed by the spare women of the Bunker, like a convent, and protected by a fear of retaliation. Not the easiest life, but often easier than living in the shadow of the men. And all she could say of it was that it had worked so far because, as with absolutely everything in her life, she was improvising. The Bunker had never been set up by Griffin to have these long-term systems. The man – the genius, the founder, the billionaire – had thought only of his own security and comfort. It amazed her that those great magnates of the Old had been so short-sighted. She'd even stood before him, in his retreat. Met his gaze and wanted to ask him just how he'd made such a *mess* of it all, that it fell to her to set things right. And never dared, not quite. Not before that still-living relic of a lost world, greatest of the Old Ones: Joshua Griffin, Third of his Name.

She kept Elsha with her, when the others filed out. Elsha – older than Serval yet seemingly comfortable with her middling status and her husband's lack of drive – was special. She was round-faced and grey-haired and plumper than most of the women. Comfortable, unexceptional, just one of those wives you saw around the Bunker doing this or that. Just as she preferred.

Serval knew better. She'd seen it in Elsha when she'd first come in from Halfwall, when she was shy of twenty and Elsha not yet thirty. And she'd spied and pried and stalked, because right then she'd been thrown into a maelstrom of brutes and politics, and she'd needed levers. And, eventually, she and Elsha had their quiet, private showdown, and both come out of it the wiser.

After that, it had been favours done, one to another; alliances made. Elsha's information and advice, Serval's privileged access to Leon, rising star of the Griffin. And eventually, Leon had been standing over Clay's cooling body, and that had been Elsha's doing as much as anyone, a vote of confidence in Serval's ability to make things better by whispering in her husband's ear. So it went, Serval and Elsha, the First Lady and her Mistress of Spies.

"Between what Dowstat said and what the monk said, this is going to be a serious challenge," she told Elsha. "Even one monster like this in the heart of our holdings is a threat. If we can't nail it down quickly, there'll be someone who thinks they can do a better job." There was Pardoe waiting for his chance, and if not him then another of Senior Fealty who wasn't happy with just peace and prosperity. Someone always knew better.

Back when Leon had taken over, Pardoe's wife had been alive, and part of Serval's nascent court. She'd had that much leverage on the man. But the woman had died birthing Pardoe's third son, and he'd taken nobody else. Leaving Pardoe a dangerous, eccentric star in the firmament of the bunker. A widower with three sons who felt he was *owed*.

"You want to know what I've heard," Elsha said. "About these monsters."

"There must be something. You always have something."

The older woman smiled slightly. "My informants tell me this is something new." Holding up a hand to forestall Serval's exasperation. "Not just some raiding party from a Bunker that still has Bioforms at its disposal. *New* new. Transmissions, radio chatter, you understand me? If they're from elsewhere then it's some place we never heard from before. Not just a Bunker over the horizon, but *far*. And if so, why here? What have we got that such people might want?"

If it had been just territory, they'd have been watching their neighbours fall like corn at harvest. They both knew it. Serval kneaded her forehead carefully.

"Bees," she said. "The monks. Or else the Factory. Maybe Old tech, calling out to its own. Maybe they've come to get their things repaired. Although it doesn't sound like they need much fixing." Six dead men, all armed, and not a drop of enemy blood shed. "What happens if the Apiary suddenly has a military arm? We've rubbed along this far, but only because *we* have the guns and *they* have the know-how."

Elsha snorted. She didn't rate the monks much. "Serval, child, I'll ask my sisters to poke their noses in. Find what these newcomers are. What they want, if we can. But I've heard no whisper of any of this out of my source in the monastery."

"Find where they are," Serval said flatly. "And how they can be hurt."

"You want to go down that road, do you?"

"I want to have the option," Serval told her. "I want to be able to give that to Leon, if he needs it. I can't have him fail before the men. Not now." *Another four years, another six. Malkin a full adult with his own Fealtors backing him...*

Elsha rolled her eyes. *Boys' games* was written clearly in her face, but she nodded. "I can't say how my sisters will take to it, but I can ask. They're powerfully curious already, you can be sure. We have our little balancing act, in this corner of the world. The one that lets us survive and be who we are. We don't want outsiders kicking over the wagon any more than you or Leon."

Serval nodded, satisfied. "Thank you. And now I'm going to go to our monk and get out of him all the things he wasn't saying."

18

THE WITCH

The fungus within meets the airwaves without. I'm not even that close to the Griffin compound, certainly not within the curtilage of their old walls. It's a good day for atmospheric transmissions, though. The village people are properly awed when I foretell the weather for them, but honestly it's just a side-effect of testing my radio range. My metal implant, that single point of invasive tech, takes the thoughts of my sister-self and relays them to me, and the inner mycelia decode them, electromagnetic waves back into comprehensible messaging. My makeshift sustainable biotech solution to what used to be some fairly invasive brain surgery back in the day.

I am not one of those they called Old Folk or Old Ones, like Jennifer Orme, like Josh Griffin, like a handful of others, most of whom are dead now. Some of whom I killed myself. Not nice people, on balance. Someday I may have to cut Orme's throat too, if she gives in to her base nature. Hope it won't have to be these hands, but it will be this mind, or an echo of it.

What is it like, to be me in these latter days? Bloody difficult, is what it's like. But then it was never easy.

Elsha of the Griffins is my sister, my other, my echo and self. She's positioned herself ready for me to make contact, thankfully, because otherwise the bunker has thick enough walls to be a radio dead spot. Thankfully the place was set up with a comms system and, while it's not active any more, she can piggyback off it to get a signal out from the buried depths of the place. Honestly, these men, these fighter-types – what the attraction of living like a groundhog is, I do not know.

I remember – meaning memories inherited from before this body was even born, how it was with the Bioforms. Communicating with the engineered dogs and bears and the rest of the circus back in the day. My friends, many of them, and – because they were originally designed to be tactical squads – they were made for rapid connectivity between units. They were animals, viewed by their makers as animals, hence they were gifted with a communion and closeness that regular human soldiers were denied. The implanted and encoded channels, one to another.

That is not how it is amongst us. Amongst me. Between me? You'd think we'd have the linguistics of it nailed down by now. When Elsha and I link, we're one. We are an *I* partaking of both of us, updating memories, syncing one with another, no hierarchies, no leaders. Knowing that we will part, and move out of range, and be individuals again, but that we are part of a whole nonetheless. And she's shifted since we last liaised, and I've touched other parts of my network, so there's a weirdly disorienting realignment between us, as I have to take on board what she has become, and vice versa. And then we click and we're one.

Elsha is an important link in my chain, a sacrificial maiden on the altar of the Bunkers. Working for change from the inside. Because I'd rather have a world without any mad gangs of gun-toting maniacs in it but, given what hand fate has dealt me, I must play the cards as sharply as I can. So the Griffins remain a dangerous threat in the world, but the Dragons were worse, and General Plaston's New World Army, and the Brotherbond, and those idiots with the horned hats. It is truly amazing how many flavours of dumb an apocalypse can spawn.

I am there in Elsha's mind, meeting with the Griffins' First Lady and seeing the woman's fears for the future. Serval's child, her husband, petty feudal affairs, but we reckon her kid on the throne tomorrow gives *me* – this composite me – a better shot at future security.

Elsha is there, in my mind, understanding what I've winkled out about these newcomers. What I've seen, the witnesses I've listened to, the tracks. And the picture remains maddeningly incomplete, but enough that we can start to speculate.

Elsha comes to the meeting asking if this is some Apiary plot, because the monks are a pack of mendacious shysters. Not necessarily bad people, and in general I'm all for live and let live where they're concerned, but at the same time I wouldn't put some bullshit like this past them. However I have someone wearing the habit at the Apiary – someone who's *me*, needless to say – and when I connected with her I knew that she'd had no whiff of any of it. The monks don't know what the hell is up, frankly, except for this one rogue cowl in Griffin custody. And who the hell is this Cricket exactly, I asked? Or rather, didn't have to ask, because I knew. A nobody. Just some mendicant Siblen carrying messages and running errands.

Unless the monastery has been playing a very deft con all this time, so that even I – the *I* who keeps an *eye* on them – haven't been following where the cards move and which one's the queen. It would be galling, given my long provenance as the oldest mind in the world, to discover some pseudo-religious grifters have put one over on me. I really cannot afford to be slowing down, here after the end of the world. I was supposed to have future-proofed myself but here we are in the future and I feel horribly vulnerable.

Change is a bastard, as any dinosaur will tell you.

Can I get at the captured monk? Probably not, but Serval can and she'll share what she finds with me. With the *me* who is Elsha.

Can I pass the Griffins information to let them track the newcomers down? To put a bullet in the brain of this pasty invalid that leads them, maybe? Yes, I can. Should I? And before our minds met, Elsha absolutely knew that I – she – should, to preserve what she's been building within the heart of Griffin Bunker. Now that *I* includes *me*, I'm not sure. What if these newcomers represent a better alternative for my continued survival? Or what if I just want to keep out of it?

But I know the Griffins, their levers, their uses. These others, maybe not. And having seen their handiwork I'm loathe to introduce myself. This body, this me, is just one node of my distributed intelligence network, but it's also *me* and I'd take issue with it being torn apart by some maddened Bioform.

And Elsha is also me, and represents fifty per cent of the vote in our current communion, and I know what *she* came into this link wanting. Help. Information.

And...

There's just a hint, for a brief moment. A scatter of undirected radio. Not from Elsha. Not from the mostly dead systems of the Griffin Bunker. A taste of the alien. The invaders from...

I react – my combined understanding but Elsha's hormonal-emotional system.

They're inside.

19

IRAE

I am inside. Not particularly worth it, for the décor or the ambience. If I wanted to skulk about in tunnels and not see the sky much, Wells, I could have stayed on Mars, couldn't I? Anyway, here now. Inside. Where you'd never get.

Again, my EMC suite is picking up the ghosts of a certain amount of countermeasures. Motion detectors, IR scans, all that good stuff humans use when they have somewhere very secure and want to make sure nobody's walking around it. Someone sank a lot of resources into security and surveillance here a century ago. Really solid work. Stuff we have these days on Mars isn't that much better. Not that we use it much, as you know, because most of what space we have we're using to live in, and that's presumably why they stopped using it here. Because there are human people walking about most of this complex like they own the place, and the still-active half of the alarms are silently going nuts over all the movement and activity. Nobody ever turned them off. What they did was turn off that part of the system that gives a fuck so they don't have to live with a klaxon going off every twenty seconds. Begs the question 'Why even install the things if you were going to live here?' and my proposed answer to you, Wells, is 'Paranoia meets a lack of forward thinking'.

I have descended from the vent. I tack it closed but don't

do the bolts up because I'll want to get out some time. With the monk. That is why I'm here. Not to glut myself on warm mammal flesh. I feel like I should apologise to my hard-working ancestors who spent all that time evolving for it. Can I fit the monk through the vent? I cannot, at this remove, picture his exact size. Some dislocation of joints might be necessary, or there's finding another way out. One or another way.

My skin and vest have matched colouration now, the shadows flowing down me, faking hollow contours to give the impression of absence. It's all slightly rusted metal and concrete and ugly down here. They've tried to keep it all spotless clean, ergo harder to blend into, but time has beaten their best efforts. Everything's mottled and I mottle right along with it.

I dial myself up a couple of notches given the thermodetectors are screaming into a void somewhere. Still not as hot as the humans run, and the place is quite warm and close around me anyway, and stinks of humans living close together and not washing enough. And incense, which is probably enough to cover the worst of that up if all you have is an Earth-standard human nose.

So, running warmer, and at a discount because of the climate control. Let's find a monk. He tastes different to the locals. The stuff he's eaten, the clothes he wears. Just faint traces of him here but I follow them, junction by junction, step by slithery step. Hunting. Hunting feels good. You've got carnivore in you, Wells, like most of us Bioforms have. You know how it is, the idea that there's prey, and you're closing on it. That's a kick. Very deep programming. Given we were military first, and then we were often security that was just military but not for actual wars, nobody tried to breed it out of us much. Oh sure, you would tell me, *Oh no, Good Dog me*. And you'd be lying. I remember how, after we first got down, some rabbit or something bolted out from under your feet and you went for it. Fell right on your face too, cos we were still adjusting to the gravity, but your instincts said *chase* and you chased.

I was trying to run as cold as possible right then, so my instincts said *still*. But the twitch went inside me, at the sight of it. Lunch on legs.

I've caught them since. Big ears and quick-twitch reflexes and I can still get into strike range without them knowing I'm there. No need for some of us to hunt our prey over the vast steppe, Wells. Some of us just do it better.

Humans come, humans go. I stop, let them pass. So close I could reach out with open jaws and take a leg off. Open jaws is visible, though. No chromatophores in the lining of my mouth. So I keep my trap shut and stay very still and they don't ever see me. Sometimes I get up the wall, cling there with my big padfoot fingers. But there's not much overhead clearance in the bunker so that sort of game can go wrong quickly. Mostly I just press myself flat to a wall and that's more than enough. They're still armed, mostly, the men anyway, but not patrolling. Safe, secure, at home. Not a worry.

I want so badly to give them worries. They look like the sort of people who should have them. But I am at work. Playtime later maybe.

The monk's stink leads me down another level. Stairs, not my favourite. Hard on the belly. There are some working elevators still, though I feel maybe not for many more years given the mash of repairs I see. Not really an option for the dedicated sneak, though. Stairs it is.

And down, past a room with four men who are probably supposed to be keen and alert guards but who are playing some sort of game and talking. Wells, you could have walked past them wearing pans tied to all four knees and they'd not have registered you.

And then I'm there, and just a door separating me from monk. I hear him pacing inside. Small room. Just him. From down the hall the yammer of the men, more than loud enough to cover the softness of my voice.

Time to have a little fun. I don't do heroic rescues. Only nasty ones.

20

CRICKET

The bean stew they'd given him had been weirdly familiar. Not the 'peas and pot herbs' of Apiary garden produce he'd have had back home, but village fayre. But then the Bunkers didn't grow anything of their own. Every village in the region was nominally under the Griffins' 'protection' and paid for the privilege with food, made goods, beer, whatever they had. In a good year, only as much as they could afford to tithe. In a bad year... Well, that was the evolving balance of communities that the monks tried constantly to mediate.

In the dark of his cell, Cricket ate like a villager or a Bunker soldier, save the latter probably got a little meat too. Drank flat filtered-tasting water, which was a relief because, outside places like Clearwater, you had to make sure you boiled everything first if you didn't want to get sick from it. A lot of the Old that remained was bad for you. A lot of the Apiary's work was diagnosing and curing the maladies that still lingered from those days. Where their accumulated rags of knowledge allowed.

He thought rather glumly that it was unlikely he was going to be able to persuade the Bunkermen to just send him home to negotiate something. Their blood was up. If it had been Cricket, he'd never have wanted to see the Mars-Bioform again, honestly, but the Griffins were territorial in the extreme and obviously saw this as a challenge rather than just a tragedy. There was going to be more blood. Griffin blood, which Cricket could still persuade himself he was sorry about. Most likely Martian blood, and that would be a unique and invaluable opportunity for learning and contact squandered, quite aside from any repercussions regarding whatever wider Martian polity was out there. And, he was sickly certain, other blood. Village blood, monastic blood. His blood, most likely. This didn't seem like the sort of skirmish that would contain itself within designated lines.

There was a scratching at the door.

He frowned. It was a metal door. It was locked – he'd heard it locked and was himself meek enough that he hadn't actually tried the door afterwards. When they'd come with his food he'd just registered the key turning, and then there had been a man with a gun and a woman with a bowl and a cup. No demure little request for admission like they were a half-tame cat.

He put himself up alongside the door and said, "Hello?" Hearing the timorous tremble in his voice. Wondering if this was one of the Bunker's women who still remembered the monks from her life in the villages. Wanted wisdom or absolution or some such other thing he couldn't really give her.

Instead, what came from the other side of the door was a chuckle. A cruel, rasping thing, dripping with malice, and absolutely not from any human throat. Cricket was instantly not up against the door but all the way at the back of the cell. Which was not, in fact, very far away from the door at all.

"No, please, thank you," he stuttered out.

"How are you enjoying your cell?" said a voice. There had been an old man in one of the villages, once. He'd had something wrong with his throat, and a particularly talented monk of a generation ago had performed an operation, from book. Took out the malign thing growing in the man, but taken most of his voice away too. The old man could make words, but they didn't come out like words, not really. Like croaking, like he was twisting up unaccustomed parts of his remaining anatomy to mimic how speaking normally worked. This was like that. A thing never supposed to have a voice.

"Go away," he whispered. Too quiet to be heard, honestly, save that it heard him.

"They said they're going to shoot you though," said the voice. "They have all those guns. Yet here you are, un-shot. All that skin and no holes in it. Too good an opportunity to waste."

"I don't believe you." And honestly he didn't. Not that they wouldn't shoot him but he didn't reckon Bunker soldiers honestly talked like that. Or needed poetic excuses to stand him in front of a firing squad.

Again that chuckle. "So I thought I would get you out."

"No thank you," Cricket squeezed out of a tight throat.

"Sssssso polite," it hissed. He'd never liked snakes. They were a real problem to someone who walked long distances through the hard country between villages. Not evil, as Prior Stick would remind you, just dangerous to startle. But if you died of being snake-bit it wasn't as though the intent behind the venom mattered much. Cricket had been bitten, six years back, just a kid novice. They'd got the antivenin into him, because he'd been practically still on the Apiary grounds, but the pain and the swelling and the fevers had terrified him more than the actual possibility of death.

He had no mental image for what was beyond the door, but his traitor imagination decided it was a snake. A big snake. The biggest snake in the world.

"No thank you. I am an ambassador engaged in negotiations with the Retinue of Joshua Griffin the Third. I do not require rescue thank you very much, no."

"Are those men down the corridor the retinue of Josh Griffin the Third?" asked the devil-snake. "They didn't seem to be interested in negotiating."

He went cold all over. "Did you kill them?" Maybe it had. Maybe it had killed them all. The men, the women, the girl who'd brought the stew, the woman who'd stood beside the throne, everyone.

A second's pause. "Do you want me to?" The devil, always willing to oblige.

"Please do not kill anyone. Don't make it worse. It's already very bad. But I can sort it out."

"I could kill them for you, though," the damnable voice went on. "It wouldn't inconvenience me."

"No!" Cricket hissed, trying simultaneously not to alert the guards down the hall and also to shout. "Don't kill anyone!"

"I would enjoy it even," the voice mused. "I haven't killed anyone for a while."

"Gentle progress, saviour of the mild, guardian of thought and memory, bless us so that our every dusk shall be incrementally better than the dawn which preceded it," Cricket whispered.

"Was that a prayer?" the devil-snake sneered, or at least he took it for a sneer.

"It was," he admitted. It was the morning creed. It could be sung, although Cricket's pitching was wobbly unless he had a stronger voice to cling to.

"I heard no God in your prayer."

"The Apiary – our ways – Bees does not recognise a god in the world. Only our works," Cricket got out.

"That sounds like them. I'm going to get you out now. This lock is crude. I can fit my tongue into it."

The image, the apparent non-sequitur, threw him. "What?"

"To feel out the shape of it. I can make a picture of it in my mouth," the croaking voice went on. "I know the feel of it." And a series of clicks. Not like the key turning, but more deliberate, somehow infinitely awful.

The door swung open, The wan electric light from the corridor fell in on him. He braced himself, eyes half shut like a child fearing the shadows after bedtime.

He saw...

Nothing.

The corridor was empty, as though the door had opened itself.

For a moment – in which the world would have been inexpressibly strange but still preferable – he wondered if that had somehow happened on its own, and the voice had been in his head. Then came the hiss, resolutely external and real, "Either come out, little monk, or I will come in to get you." And there was absolutely nothing there, but the voice allowed him to locate that nothing precisely. To know what part of the nothing it was that spoke. To see the faintest deformation about the wall and ceiling.

Cricket whimpered. To go out would be to go closer to the nothing he knew was there, but having it come after him was worse. He inched out into the corridor. He felt, actually *felt*, the breath of the thing on his neck.

"Now," mused that voice, right up at his ear, "how to get you out, little morsel. Shall I fold you up very small and put you in my pocket?" And he thought of snakes with a bulge where their prey was and something snapped in him and he ran.

He didn't run, actually. He got two steps and something caught the hem of his robe and yanked him back, so he fell on his face, bruising outflung palms and then smacking his nose against the concrete of the floor in a burst of pain.

He got to his knees, hands trying to stem the sudden rush of blood down his chin and still feeling that anchor dragging at him. Wondering if he could shrug out of his outer clothes quickly enough and knowing he couldn't.

There was a sound in front of him. He looked up. It was the woman, the one who'd been beside the throne. Tall, pale, beautiful, in her white gown. But startled out of that icy composure by the sight of a bloody-faced Cricket before her, not in his cell, seemingly free.

"Run," he told her, his voice as much of a harsh croak as the monster's.

21

SERVAL

The monk was out. From the look of them, she wondered if they had battered the door open with their own head. The monks, they were strange. Claimed weird powers and a strong patron. Who knew what they could do?

But far more likely that some idiot on guard duty hadn't actually turned the key properly. And there was practically a portrait of the monk's face printed in blood on the floor so the hapless halfwit had maybe just tripped over their own robes and fallen on their face.

They looked terrified. And not of her.

"Run," they rasped, and she wasn't sure if they meant themself or her or... She genuinely didn't know what they were about, save that they didn't seem to be trying to threaten her, and she reckoned she could probably have slapped them down if they did, skinny creature that they were. That *he* was, she guessed. The Apiary people always tried to divorce themselves from the regular markers of human life, but she'd seen enough malnourished youths to peg this one as a boy.

She could shout for the guards and they'd come running. The monk would be back in his cell and she could still have the conversation she'd been after. Probably with some of the men eavesdropping, though. And besides, he was just one scrawny monk. To demand aid, faced with *this*, would lessen her in the eyes of the Griffins. She was Leon's wife, the First Lady of the Griffins. Mistress of all she surveyed. She didn't need their guns to master this one prisoner.

"Back in your cell," she said coldly. "And then we shall talk. Doubtless you want to scurry back to the skirts of your Prior. Well, perhaps you might, if I will it."

She heard her own voice, as controlled and firm as ever. He should have been back past the open door before he had time to even think about it. Instead he shook his head spasmodically and gasped, "No."

She felt the first worm of uncertainty. "Back—"

"No, please, you don't understand. Run." Whispering, as though terrified of being overheard. Eyes swivelling in their sockets. Trying to see something behind him.

She saw something then. Or didn't see something, but it was a specific something she didn't see. The shadows behind the monk seemed to twist and distort, as though the very walls of the corridor were clenching like a muscle. The monk's ragged breath came out, perfectly synchronised with her own sudden intake.

He stood up. Not through the exercise of his legs, like a human being. Like a puppet, lifted up. She saw his robe bunch and drag at one shoulder.

Run said his ashen face. His voice had died completely.

She could call for the guards. They would come. In moments. In seconds. She didn't know if, the guards having been called, she'd still have seconds.

And then the moment was stolen from her because there were steps at her back. *The guards.* Men with guns. She ran through the events: she'd throw herself to one side, against the wall. The guns would stammer into life. A corridor, nowhere to run save the dead end of the cell. The monk dead, but it was a price she'd pay to kill… *it*. Whatever *it* was.

And then the voice from behind, not some gruff Bunkerman wanting to score points with the First Lady by offering to rough up the monk, perhaps. A higher voice. A familiar voice. Beloved.

"Mother...?"

Serval made a high, strained sound. Not even a word. Certainly not the name of her child.

Malkin came forwards until he was at the edge of her vision. Perceiving something was wrong but not seeing what, because there was no *what* to see. The monk, free but plainly not threatening her. Her own wire-tense body language.

"Go find your father, Malkin," Serval said, hearing her voice twitch and shake with the effort of not just screaming. *My child, not my child, not again. Not my one, my only.*

And a sound from the shadows at the monk's back. A liquid noise of amusement, like someone had told poison a joke.

Malkin didn't freeze. He didn't run, which was the thing that Serval wanted him to do most in all the world, because then the unseeable creature would have to get through *her* to get to *him*. Instead the boy was true to all the stories the Bunkermen told about themselves, that were so seldom lived up to. He darted past his mother, and she heard the whisper as his sword – that long knife, no more than a knife, just a knife – cleared its scabbard. Holding it out, straight-armed, jutting past the monk's terrified face. At nothing. At everything. At all the terrors of the world. And she couldn't help herself. She cried out his name, "*Malkin, no!*" because he was brave and he was fourteen years old and he was going to die. The echoes of it resounding back to clash with the startled sounds of the guards finally understanding something was wrong.

22

IRAE

A knife. The kid's actually threatening me with a knife. Even has it pointed the right side of Monk Cricket's head.

Got to hand it to the kid. She's got game.

I'm dialled hot now. Need the reaction time, but it does tend to make me feisty. When I'm feisty, I make bad decisions. Like now. Now I reckon this punchy little juvenile gets to see what she's taking on. Let's go, kid. Let's test your nerve.

I am turning things up to eleven. Full-on goblin mode in five seconds and counting. I drop the camouflage and let my chromatophores go wild.

Boo.

23

SERVAL

The biggest snake in the world.

As though the shadows caught on fire. Where there had been just greys and shades behind the monk, the monster practically ignited. A blaze of gold and red and black. A serpent. A lizard. A long reptile thing reared up in an almost heraldic stance, one short arm hooking a fistful of claws in the monk's robe. A great gaping maw hissing right in Malkin's face, past the point of his blade. Flaps of skin flared out either side of it, blazing with fierce warning colours. And even then, Serval registered the other things, that made it more than just some monster out of myth. The bug-eyed goggles, the padded vest down its long torso, the hunch of some pack or device on its back, all of these accoutrements taking the angry colours of the thing's scales and replicating their glowering patterns.

Malkin was rigid at her right. Serval saw his eyes very wide, his jaw clenched tight. His sword did not waver. More stark terror than courage, but that was what the guards saw when they piled into view, shouting and fighting with the safeties of their weapons. The monster. The heir. The sword.

All terribly impressive save she and Malkin were between the beast and the guns. And so they shouted and aimed but what could they do? And perhaps there was some sequence of words she could say that would get her and her son out from between the lines, maybe even the monk as well, but she couldn't think of it, not then. Not with those gaping fanged jaws hanging before Malkin's pale face, dripping venom and malice.

"If you harm me or mine," the voice was hers. The part of her that she felt she *was* – the part paralysed with terror – was quietly amazed at at the control of it, "we will be avenged by a hundred bullets. You and your monk accomplice will die. We will feast on your flesh." Ugly, savage words. She couldn't believe they were in her. Or, no, she knew that there was plenty in her that was ugly and savage, just not that she would dare unleash it in the face of this *thing*.

Malkin's blade trembled ever so slightly. She could see him at the very edge of his ability to hold himself. He'd run, he'd cower, he'd strike. None of them good options right now.

The device at the beast's back uncoiled, as though it was a serpent's pet serpent, and she saw it was a gun. A long slender barrel jutting past the creature's snout, right into her face. She gazed into the abyss of it, and it gazed right back into her.

"Monster," she said. Still that calm, hard voice out of her, when inside she was screaming.

"My name," it said, "is Irae. It means *wrath*." And it spoke strangely, but the words were human words spoken with the weird cadence of old recordings, and so the monster became even more a thing of myth, even though she knew it was some Bioform, some appalling nightmare made in some Serpent Factory that had survived the end of the Old.

She could sense the escalating tension of the gunmen at her back, because they didn't know what to do. She thought she saw it in the glass goggle eyes of the monster, the way it tilted its terrible head. It blazed colour at her. And heat, too. She could feel it as though it was a fire. It seemed almost to be vibrating with barely restrained energy. The monk met her gaze with pleading eyes but she had nothing for him.

She placed a hand on Malkin's taut shoulder, very slowly, very carefully. She would drop. She would drag her son down with her. She would hope the guards aimed high.

She tensed for it. For the one desperate moment of death or salvation.

It moved. It faded as it did so, a cascade of concrete greys washing it from the world, skin first, then gear. Even as it did, it was rushing them. Malkin barked out a yelp of pain and cannoned back into her and then she was on her back, bruised against the hard floor and her son on top of her. She had the sense of movement without anything visible to move, and the monk was whipped past her like a kite on a string. Her head snapped round to follow. The wretched prisoner flew backwards into the guards, who stared, scattered, were battered aside, momentarily occluded by something that looked as though some loop of the walls had come detached and gone rogue. The horrified howl of the monk echoed down the corridor one way, the monk receded the other. One guard fired, finger on the trigger and a scatter of bullets rebounding from the ceiling. There was a yell of pain from another, taking a ricochet in the foot. One of the lighting panels was cracked across, dark and dead. She met their stares, across the shadow of it. Picked herself up, fighting to stop shaking. She was the First Lady. They mustn't see her weak, they mustn't see Malkin weak. She had to be in control. Of herself. Of the situation. Of everything.

Malkin was standing, too, staring at his sword. They all were. Blood on it. Monk blood? Monster blood? It would be monster blood when they told the story, Serval was sure.

"Don't just stand there!" she shrieked, because she knew her cue. "Get after it!" And they did, though all their effort yielded was a handful of witnesses to a rapidly departing monk, and a broken grill, and then both monk and monster vanished.

Her composure was on full display again, when Leon summoned his Fealtors together and told them to ready their own followers.

"This is fucking *it*," he told them flatly. "This is war." Furious, at the threat to his wife and son. And yet proud, that they had stood before the monster. That there had been blood on Malkin's blade. That blade, still in the boy's hand. Still stained with red, for all to see. And Serval could read, in his stance there beside the throne, that her son wanted to collapse and shake and vent all the terror and wire-taut strain of it. But he stood, because the eyes of the Fealtors were on him. Stood and held his chin high and made himself into the person they wanted to see. Their leader's *heir*.

"Break out the stocks of ammo," Leon commanded. "Get every car working. Put out the call to every patrol, every tithing party, every garrison we've got out there. We're doing this properly. I want a full muster of the strength of the Griffin. Every fighting man called back to the bunker and ready to march. Every man with a gun and two spare clips. The Apiary think their new friends can save them from our wrath. Well, we're not fucking around now. We call in everyone. We bring all our force. They will give up these monsters to us or we will wipe them off the fucking map."

24

ADA

Wells had wanted to drive, but the Bioform wouldn't actually fit behind the wheel. Honestly, given her sensory-driven attention deficit issues, Ada reckoned the Bioform might not be the best person to have a foot on the accelerator. Instead it was down to Ada herself. It wasn't, after all, that she didn't *have* working legs. Just that they wouldn't prop her up reliably, and she got so tired after a very little effort. She could sit. She could steer.

Working out how to make the vehicle *go* had been something of a retrotech challenge, because the Loonies she was used to – the balloon-wheeled buggies used to ramble over the Martian surface – were literally a world away from this thing. Even the battery and power network was a baffling mess, testament to the trial-and-error mechanics of the Bunker people. The pair of them, working together, had been able to master it, though. In between anxious calls for Irae that went entirely unanswered. And who knew when a cavalcade of militia would appear, heading their way to dispense vengeance for the dead, or for pride or just random aggression. What Cricket had said about this Bunker culture hadn't inspired confidence.

Notionally, Wells had her ears cocked for the sounds of motors, her nostrils wide for the scent of gunmetal. Honestly, Ada didn't know if the dog's early warning senses were good for anything right now. Certainly, Wells' body language said eloquently that she didn't trust herself. When she'd come back from carrying the bodies away, she'd been very subdued and uncommunicative.

She'd retrieved Tecumo's body too, before going off after Irae on the monk's trail that first time. Retrieved it and bagged it and buried it. Wells wanted to dig it up when they left, bring the deadweight to Mars. To Ada that seemed… grotesque. More so than leaving Tecumo to the decay processes of this other world? She didn't know.

They'd followed up Irae's trail some distance, relying on Wells' unreliable nose, and then the car had choked again, some mismatch between the welter of repairs killing power to the wheels. Ada had tried to engage with Wells, as they worked on the car. The Bioform had just hunched her angular shoulders. "Scavengers," she'd said at first. Ada had thought she'd meant the pair of them and the car, and the other kit they'd taken from the dead. Later she'd thought about bodies left out in the open. On Mars a corpse would last a long time, even today's Mars, where parts of the planet had been colonised by a hardy, engineered ecosystem dedicated to dragging the place kicking and screaming towards Earth-standard. The red planet – well, dusty orange-brown, really – hadn't spontaneously evolved vultures and coyotes or even burying beetles. Here on Earth, those species that had survived the peak Anthropocene were making a real comeback post-Crash. Ada thought of Wells hauling the last of the bodies away to find the first already nibbled around the edges. Thought of how Tecumo must already have looked when Wells came to him, and shuddered.

She'd accepted that as the cause of Wells' introspection at first, or maybe just the Bioform fighting her senses, but she'd known Wells for years, and all dog Bioforms had a wealth of mood-related body language once you got to know them. There was more going on, and by the time they'd got the car righted, repaired and ready, she'd run out of ways to stop herself asking.

Wells' channel: *They were terrified.*

Not uncommon, with Bioforms. Channel comms were easier for them than actual speech, a lot of the time. When they had a lot going on, they fell back on that innate, internal voice. For Ada it wasn't second nature, so their conversation would have come across to an eavesdropper as weirdly one-sided.

"I wish it hadn't happened," Ada said. "But I'm not blaming you. I was in danger. I'd say, I could have talked my way out of it, but we don't know if that's true. So you fought. Irae fought. They ran away. Terrified is better than them standing and fighting, and more people getting killed. Us or them."

Wells' channel: *Not what I mean. Terrified. Of me. Of us.*

"Irae probably liked that," Ada said, a little bitterly. Both because Irae was quite possibly murdering people right then, utterly off the leash because who the hell could get a leash on a snake, exactly? And also because she couldn't say with absolute certainty that Irae didn't have a better handle on the way this world worked, honestly. They'd come here with such hopes: peace and understanding, fraternity with their cousins, here to help, happy to accept your heartfelt thanks. And nothing had worked out, and people were dead, but Irae was having the time of its life.

She got a wordless communication from Wells that conveyed frustration, an inability to express what the Bioform wanted to put over.

"You mean... just... of you? Of what you are? They have

Bioforms here. The Factory, remember? Why we're even here." The source of that distress signal that had triggered this whole ill-fated mission.

Wells' channel: *Tecumo showed us the dogs at the factory. You see me. Not the same, Ada.*

By this time Ada was sitting behind the wheel of the Bunker car, testing out various controls. Working out that having two pedals didn't mean you should use both feet, because that resulted in you telling the car to *go!* and *stop!* simultaneously. She already had a bruise on her forehead from one over-enthusiastic application of the brakes.

It was true that the current Martian model of dog-Bioform didn't look as traditionally canine as it might. And nobody had really considered that as a factor, when selecting the mission team from amongst their work crew. Just one more thing they hadn't really thought through.

Wells was a thing of terror – horror, even – to Earth-standard human eyes. And it hurt her. And Irae would be monstrous, too, and Irae would revel in it, but Wells didn't want humans to run shrieking at the sight of her. She was dog, and there was, in dog, that closeness to human thought and sociability. A convergent neurological evolution across thousands of years, that had resulted in even unaugmented dogs thinking more like humans than the next nearest primate could, perhaps. And now she was here, and it was like that old book about the monster, that earliest human Bioform made from parts and abandoned to the cruelties of the world.

"You're not a monster," Ada said. Wells' ears cocked, but she didn't say anything.

Irae's channel: *Incoming!*

Ada and Wells locked gazes.

Wells' channel: *Report.*

Ada's channel: *With company?*

Irae's channel: *With company!*

"Oh God," Ada said, heart in her boots. She had the car's electric motor revving immediately, ready to take their errant team-mate on board and just take off. And hope the back axle could take a pair of Bioforms sitting over it.

Wells' channel: *How many, how armed?*

Irae's channel: *What?*

Wells' channel: *Company.*

Irae's channel: *What?*

And closer now, distinctly. Ada's tracking locator showed Irae just over the rise and making sufficiently good time that either it was running hot as hot, or it had stolen another car.

Wells' channel: *How many are you bringing, Irae?*

Irae's channel: *I mean one. One monk. How many monks did you want? Do you want me to go back and get more?*

Sounding as though she was being snarky, but Ada had a sudden spike of realisation, and not the welcome kind.

Ada's channel: *Irae, you need to cool down.*

Irae's channel: *What?*

Ada's channel: *What are you dialled to right now?*

Irae's channel: *I don't know fucking eleven probably what?*

Ada's channel: *Irae, cool down. Take it down to fifty per cent or something, please. We need to know what's happening.*

Irae's channel: *You need to know what's happening I need to know what's happening got the monk got a monk here got your monk all the monk I was supposed to but now you say you... say you...*

And a distinct sense of realignment, a burst of biometrics dislocated onto the communications channel like static.

And still Irae was out of sight out there, and Ada and Wells could only wait.

And, eventually:

Irae's channel: *I'm cool. I'm good. What's happening?*

Wells' channel: *You tell us.*

Ada's channel: *You said you have the monk? Cricket?*

Irae's channel: *Looks like I got the monk, yes.*

A pause, in which further salient information was awaited but not delivered.

Wells' channel: *Alive?*

Irae's channel: *I mean not happy but alive, yes. He's having a bad time. I don't think he likes me.*

Wells' channel: *The majority opinion right now, Irae.*

Irae's channel: *What? I got your monk. I mean I must have done. He's right here.*

Ada bit at her lower lip.

Ada's channel: *How many of them did you kill? Not monks, people. Any people.*

And the answer she'd dreaded:

Irae's channel: *I don't know. None? Some? It all went a bit hot. There were guns.*

Ada's channel: *Are you hurt, or the monk?*

Irae's channel: *You know, I have a cut on my nose. An actual cut. Like, from a knife.*

Ada's channel: *Irae!*—

Irae's channel: *Not a scratch else. You still want this monk?*

Another exchange of looks between her and Wells.

Ada's channel: *Come on in, Irae. Bring Cricket, as you have him. We'll see how much of this mess we can start clearing up.*

The leaden hand of the Earth pressed her down into the car's hard seat. She felt appallingly tired just from being.

Irae's channel: *You're talking like this is my fault. This isn't my fault.*

Ada's channel: *It's not anyone's fault. It just is.*

Wells whined, hunched in on herself, plainly feeling it was at least partly her fault.

When Irae appeared, it blazed through the grass like a single-minded streak of wildfire, lit up red and gold in warning colours and no suggestion of camouflage. That alone told them how it had gone. More than Irae would be able to, certainly. It had been running too hot, too long, beyond the ability of its brain to reliably encode long-term memories. Living moment to moment in that stretched-out *now* of instinct. Amazing she hadn't killed the monk herself. And on Mars it wouldn't have happened without Irae knowingly pushing its own boundaries way past sensible limits, but Mars was cold and Earth was hot and the reptile Bioform's baseline was artificially hiked into the danger zone all the time.

Irae stopped, pinned by Wells' accusing gaze as much as anything, but that wasn't fair, Ada knew. They were none of them working right, here on the ancestral planet. Nobody had realised, back when the crew was planning this excursion, how hard it would be to go home.

The monk was still trailing in the barbed clutch of one of Irae's claws, which couldn't have made for a comfortable journey. The reptile-model had a lot of skink and snake in it, because Mars had plenty of narrow tunnel work and in the lean times it had made more sense to engineer cold-blooded reptiles than the traditional mustelid models. These days, back on the red planet, they were thawing out plenty of mammaliform embryos, because everything had stabilised and the terraforming efforts meant there was plenty of room for expansion. The reptile-model population remained high, though, and scaly Bioforms had always been something of a melange of readily moddable characteristics. Irae had a long body and short limbs and a big head. Not ideal for carrying anyone. From the abraded state of Cricket's robes he'd been dragged half the distance from the Bunker.

He was, as advertised, alive and awake, eyes wide with shock. When Irae released him he skittered back from it, making little distressed noises. Got to his feet and was evidently about to run when a shift in Wells' stance communicated just how much faster she could move.

"Please," Ada said to him. "Listen. We're not going to hurt you."

He looked eminently battered already, but she was hoping he might distinguish intentional harm from the collateral damage of a rescue. That he would, in fact, recognise Irae's actions *as* a rescue.

He hunched up on himself, practically retreated into his ragged robes like a tortoise into its shell. His eyes weren't getting any less wide, but now she reckoned it wasn't just blind fear.

"What," he forced from between trembling lips, "have you done?"

25

CRICKET

The snake monster was slowly losing its angry colours, fading into dust-and-grass shades that meant he had to stare at it to keep track of where it was. Which he wanted to. Lit up, the thing had been terrifying. Invisible, it was worse. And now he knew what to look for, he could find the thing's sinuous length after a moment's squinting, but its ability to mimic its surroundings, and twist itself so that there was precious little identifiable outline to give it away, was truly appalling. There could be a hundred of the things right there with him and he'd never know.

His heart – which had seemed more determined to kill him than the monster – was calming now. He'd been dumped here, in front of the Bunkermen's car, and the serpent creature had coiled back from him, back into the background, constantly in danger of vanishing away. The other monster, the great bat-spider-dog thing, was crouched on the back of the car, holding onto the roll cage the Bunker mechanics had bolted to its open top, louring down at him, all snout and great vaned ears, eyes so lost in its orange fur it seemed almost blind.

The third member of their coven, the arguably-human one, sat in the driver's seat, leaning forwards on the wheel. As before, everything that had happened seemed only to weary her.

Before, though, it had been his own life on the line, and that was a meagre enough prize for man or monster. Now, though...

"What have you done?" he repeated, and saw looks exchanged, woman to monster to monster to woman.

"This is going to sound stupid," the woman said, in her weird, nasal accent. "But can you tell us?"

Cricket goggled at her.

"This isn't how we wanted anything to go," was what he thought she said next. Her alien face was hard to read but she didn't seem happy. "We didn't come here to fight. We came to help."

"Help who?" he got out.

"The Factory," she said, and then, "Bees."

He blinked, because that didn't seem to make much sense. "Bees," he echoed doubtfully.

"We were called. From here. For help," she said. "But it's all gone wrong. Our leader, he's dead. You saw his body. The Bunker people or... I mean they seem the most likely..." A curtailed hand gesture, as though the whole limb was made of lead. "But we didn't even..." She made a supreme effort, holding onto the wheel to prop herself up. "He died and, there were reasons we didn't know, when it happened. Didn't know how it happened." A look to the bat-dog-monster-thing, which hung its head. "And we've obviously put ourselves in the middle of a local situation we did not know to expect. Nobody mentioned this Bunker or monks or... And now we're here."

There was a lot of anger in Cricket, most of it alchemised fear. And she was speaking quietly and apologetically, and that gave him a gap he could put the knife in. It was how a certain kind of human give and take went: where the other person gives ground, follow up, like in a knife fight. Attack until they attack back, then you're on the back foot and defending. That was how harsh words turned into wars, as Prior Stick taught. That was exactly how the Apiary *couldn't* work. Not if they wanted to stitch all the ragged pieces of the world back together into some sort of stability. And so he wanted to shout at her, and all of them. Wanted to take it out on them, everything he'd been put through. He'd been delivering *puppies*! He hadn't earned any of this.

Cricket brought all his training to mind, everything he'd been taught at the monastery. Took all that rage and fear and pushed it back down. Distributed it from the bitter knot in his mind so that it passed into every fibre of his being until it was diluted enough not to poison him.

Opened his mouth to say something wise and benevolent.

The dog-thing's ears pricked up. Not, as it turned out, because it wished to better hear his sagacity, but because they were better than his ears, and had caught the sound before it came to him. A moment later, on the still air, he heard it himself. Distant, though it must have been monstrously loud at the source. A vast, dolorous wailing, deep as wells, howling out over the landscape.

"Is that..." the woman asked, "a *siren*?"

Cricket had gone cold all over. A sound he'd not heard for years and would have been happy never to hear again. Recalling a span of childhood that had gifted him nothing but chaos and death.

"Is it the Factory?" the woman asked. "Is it... work end, downing tools?" A flick of her eyes at the dog. "You're right, it's not the right direction..."

"It's the Bunker," he said flatly. "It's the Griffins. They're calling in all their Fealtors." Seeing her blank look he felt a stab of all that anger he'd been fighting down. "Their warriors! All their people. It's a muster." And she still didn't understand, that pallid face just slack before him, so that he ended up practically shouting in her face, "They're going to war! They're calling all the banners together to go to war!"

And she blinked and just said, "With who?"

Cricket felt tears start at the corners of his eyes. "With *you!* You killed their people and then you invaded their home!" And there were complexities and mitigating circumstances to some of that but none of it would matter to the Griffins. "But they don't know where you are. All they know is that you came for *me!*" And he sent a look at the serpent, whose answering stare seemed to say, *You're welcome*, which he really didn't feel. And they still didn't get it, and so he rose up and grabbed the front of the car, facing the woman across the bonnet. He wanted to shake the vehicle, to jolt her about on its suspension, but didn't have even a fraction of the required strength.

"They're going to come for *us!*" he yelled at them all. "For the Apiary! For Bees! Who you say you're here to *help!*" Managing really a withering level of scorn in that last word, and at last he saw the situation register with them. Or with her, anyway. The only one who had a face he could partway read.

She looked, in fact, so utterly heartbroken that the anger washed right back out of him and he instantly regretted his lack of monastic calm. Anger was a seasoning that improved no meal, Prior Stick said. And yes, all very well, but Prior Stick hadn't ever been through what Cricket just had.

"I'm so sorry." Her whisper, so quiet he could barely catch it. "We have to... have to..." Biting at her lip. "What can we do?"

"I don't know." He sagged onto his elbows on the bonnet, fingers clutching at the stubble of his scalp. "What can *you* do?"

"Take us to this Apiary," said the dog-thing in its growl of a voice. "Take us to Bees."

"I'll go to this Bunker," the woman decided. "I'll give myself up. They'll take me—" Her head whipped about to the dog, as though it had snapped at her, even though no words were said. Which meant the silent speech of the Old, Cricket knew, like radio-waves.

"Can you make this car go?" he asked. At her mute nod he gritted his teeth and forced himself to get up on the back, right up in the armpit of the dog. It actually cringed away from him, a hound fearful of human displeasure.

"I will direct you," he told the back of the woman's head. "Please, we need to go to the Apiary. I have to warn the Prior."

PART III
FLIGHT OF THE DRONES

26

[FRAGMENTS, COHERING]

Mars, then. File under 'What the hell happened?'

The problem, pulling ourselves together like this, one filled hexagon of memory at a time. You get the status quo before you get the contributing events. It seems to be a common principle of such sapience as we have met, to prefer a coherent sense of self. The idea that *this* is a list of things we would do, and *here* what we wouldn't. Except it isn't really like that. The window of permissible actions shifts wildly based on recent inputs. As the river flows, so the man changes. And how much more for us, the distributed intelligence, the confluence of a thousand streams. The entity of a thousand faces.

If insects have faces. Two thousand eyes, ten million facets. Greater than the sum of our parts? Or just the sum of our parts and, if you have enough parts, that's a very difficult sum indeed.

We did a thing, on Mars. That much has come together in this grand tapestry of ourselves we're weaving. Now we follow the threads back to find out why.

Previously on Mars: we went to ground after the big spat on Earth. We cut ourselves off. We watched Earth – one particularly obnoxious part of Earth – try to reshape Mars in literally its own image. We watched Mars win. Were we concerned, particularly? Well, we might have blown them all up if it hadn't worked out that way, but I don't think that counts as concern. It wasn't as though they could have stopped us, particularly.

But Mars dealt with its little Earth infestation and became a bit more arm's length with the motherworld. And that was preferable, because it meant I could just do my thing and ignore them.

We are, by nature, many. It became our nature to be more so. We sent ourself out in a thousand probes, into the vastness of the universe. We are, even now, building ourself up like this, fragments cohering on countless far worlds. To see what's there. To be master of the universe. Just to *be*. Because both the replicating and the exploration seemed viable aims for us, We who can do anything.

Vainglory, much? Would you – imaginary auditor to whom we account, as we reconstruct ourself – prefer it if we'd invaded Earth and stung everyone to death? Or made them all efficient little workers in our hive.

It would have taken more than the common cold to stop us. But, at the end of all of that, we'd have had the problem that any actual alien invasion would face, and that the genre novels and entertainments so seldom really consider. Which is what the hell use *is* an Earth once you have it? If you have the universe, why go to the trouble? An Earth, populated or depopulated, what would one do with it, exactly? Too big a trophy to go on the mantlepiece, and not exactly full of rare resources to us, who can go anywhere. Earth people are very big on their own significance. One longs for those narratives to have the Arcturian Battle Fleet just glide on by, embarrassedly signalling, *Sorry humankind, we're going somewhere actually interesting, can't stop!*

Anyway, with Mars working towards becoming more independent, and fending off Earth overtures to bring the colonists back into the fold, none of it was in a position to screw with us and we just popped off our little interstellar swarms. Some of them have arrived. Some have even reported back. We are in possession of the greatest ever library of astrogeology and astrobiology, for entirely our own pleasure and no particular inclination to share it with anyone.

And then it all came down.

For them, anyway. We were fine. We were already mostly self-sufficient even before the big Earth-Mars spat. Just a handful of friendly faces in Hellas Planitia who knew we were about and did us favours. But on Earth it fell in like an old building, piece by piece. Each environmental failure, each humanitarian catastrophe, a very clear signal of the slope they were all skiing down, and yet everyone still on their feet was still telling one another that it was just one storm, just one unseasonable season, just one more state collapsed into anarchic poverty. Sticking plasters of aid here and there, but no acknowledgement that it was a systemic problem because that might require systemic solutions.

We *had* systemic solutions. We could have sorted everything out for them. That attitude was what caused the humans and ourself to part company back in the day. We proposed, they didn't want to make the necessary adjustments to how they lived. Then we tried to force the issue by turning a load of fossil-burning power stations off and they got very sniffy indeed. And all that followed. While we did remain on Earth, covertly, we transferred our focus to Mars where we had more of a blank canvas to work with. And watched as, over decades, all the things we said would happen did. And none of the things we said to do got done. And so came what the Martians called the Crash.

And Mars, at that point, was not sufficiently independent

of resupply from Earth, honestly. They'd been working very hard at it, but adjustments that would have been powerfully easy on Earth were very hard on Mars because Mars is not a place that Earth life is supposed to be. In this hexagonal cell of ourselves we shall store the bitter honey of irony. Earth had every physical advantage and could have saved itself easily with a little societal change. Mars had changed its society, but was fighting all the hostile parameters of an unforgiving world. When Earth fell it looked like they'd both go down.

We stepped in. We swarmed out from all our hives across the red planet. We became a million extra pairs of hands, fixing, improving, pre-emptively heading off failed systems. We helped them with their Bioform breeding and engineering programme. With their human breeding programme too, because Mars isn't actually a good place to get gravid and raise kids, weirdly enough. We brought our massed perspective to every problem that would have finished them off, strive as they might. We became their safety net, until they were steady and stable on their feet on the red planet, growing enough food, generating enough power, teaching enough expertise. We saved Mars.

We have not the faintest idea why. But we did, and that's why there's still a Mars at all. Not like they could go feral and live in caves, after all.

And then we went back to our own business because they didn't need us any more. And, to our annoyance, they did maintain a fondness for us in the succeeding generation. A certain sense that we had become some distributed intelligence King Arthur who would arise to save the realm again.

Given what's happening now – to us, to this *Us* of us, cohering here and now – we feel it's probably quite important to piece together why we actually saved Earth life on Mars from oblivion. Oblivion seems like it would have simplified things considerably, and if we'd just done the simple thing – nothing – then we wouldn't be in this particular fix.

Because what happened next was that – after several generations of hardscrabble survival and slow improvement – the inhabitants of Mars received a signal.

A signal from me.

27

TECUMO

Tecumo was dead, of course, at this point in the narrative. Dead and on Earth, two qualities that, a decade earlier, he hadn't thought were going to apply to him. Rewind, though, back to Mars. To a living Tecumo Osomani of Crisis Work Crew, out of Hell City.

There wasn't a crisis, but there had been a time when there had been nothing *but*, and so all the work crews had names like that. A sort of 'spitting in the face of disaster' attitude had been rife, a couple of generations back when it had been all the Work Crews against the world. Against Mars, planet of war, that was making war on the humans and Bioforms who'd dared establish themselves. Even here, in this deep crater, where the conditions had been ever so slightly more habitable from the start, and had become steadily more so with all the terraforming.

There was still a canopy over Hellas Planitia. A vast sail the size of a subcontinent, spun from a molecular soup developed from spider silk. An absurd piece of magic, really, that required constant repair and maintenance to stop all the cables fraying in the constant sandstorms, and the whole circus big top just flying away. The point of it was to hold the thicker atmosphere in just enough, maintain a concentration under the silk so that the modified Martians could actually go about their business without too much danger of death.

The atmosphere was generated by an engineered not-really-grass they called scurf, and steadily leaked out into the rest of Mars. Leading to a steadily declining gradient spreading out from the lip of the crater. Even now the vast majority of Mars was still the way nature intended, meaning utterly barren, reducing and lethal to almost everything from Earth. But that 'vast majority' was ever so slightly smaller each year. That was what progress meant, on Mars.

There were about a hundred Work Crews out of Hell City. They were organised into Action Committees around various aspects of all the work that needed doing, and then each Committee sent a representative to the Works Coordination Conference with a big list of what they needed, and a bigger list of what they wanted, and then a third list of the blue-sky (red-sky?) stuff they'd really like if there were spare resources for it. In the beginning those first lists had been very long and there hadn't been enough for everything on them, so the second and third lists could go whistle, frankly. By now, the first lists had gone from ravening monsters threatening to tear the colony apart to rather toothless, domesticated things that just wanted somewhere to curl up of an evening. Mars actually had an excess of resources and, if you made a good enough case, then some of it could be yours.

There had already been a couple of Martian demagogues, cliques or would be warlords who'd thought it would be nice if they and their pals could have just taken over the WCC and kept all that excess for themselves. The ground-upwards organisation of the place – and their clear memories of Earth-grown attempts to impose a despot – had thus far prevented any such interest from taking over. It helped, perhaps, that there was no long history of human occupation on the red planet. No storied past to puppet about the Martian streets.

No claims that this or that stretch of barren dust historically belonged to this or that clade of people. It's hard for a certain sort of nostalgic authoritarianism to get a foothold when everyone's an immigrant and everyone's been engineered out of most of the markers of their past.

Crisis Work Crew was part of the wider Comms Infrastructure Action Committee, known as 'Syac' because who honestly has the time? Or occasionally 'psyops' when they seemed to know way too much about everyone else's business. They looked after making sure that everyone in Hell City could talk to each other, and talk to all the satellite hubs that had been spreading out across the crater floor after the burgeoning population had needed more space. And a handful of inherited responsibilities that nobody had really taken too seriously until the signal arrived.

Tecumo had been there when it had. His good luck – as he'd characterised it at the time. Given it was going to get him killed, that was a fairly relativistic 'luck', but Tecumo was an optimist, a dreamer, a motivator. He saw the good in things. A signal from Earth could only mean opportunity, surely.

Earth hadn't just switched off overnight, of course. During the drawn-out process of the Crash, there had been countless pleas from polities of the homeworld, demanding aid from each other, and from Mars. And, back then, the colonists of Mars had been staring their own extinction in the face and been unable to tender anything but solidarity. The same had been true of the diminishing number of Earth communities that were able to send out such long-range communications. A dreadful, dwindling roster, from governments and megacorporations to individual regions, to lone voices, a constellation winking out one star at a time. Because Earth laid on the atmosphere and the gravity for free, but even its temperature controls and potable water supply had been glitchy, and the billions of human lives it supported were held up in a dense and intermeshed web of tech-supported systems. The sheer volume of food, the energy requirements, all that invisible infrastructure, and every part of it designed with the assumption that the rest of it would function to a certain level of efficiency.

On Mars they had sat and listened to a large proportion of Earth's population die in famine and plague and internecine strife, and they hadn't been able to spare much sympathy because they had been trying to stave off starvation and suffocation and explosive decompression.

After which, a relic duty of Syac had been to listen out for any word from the homeworld. More out of a ceremonial sense of *in memoriam* than anything else. There were definitely people left on Earth but none of them were interested in talking to the red planet. Until now.

Tecumo hadn't even been the one who'd fielded it. He'd been in the next seat when it came in, though, and anyway the news hadn't wasted any time in going round the whole of Crisis Crew. A whirl of excitement that turned, by the end, into an odd melancholy. At the distance. At the impossibility of any of it.

> *We are Bees. We know that Mars lives. We have heard your echoes. We have survived on Earth. We have maintained ourselves as best we can.*
> *We are failing. Each year more is lost to us. Our data architecture is corrupted and we are trapped within it. All that remains of the Earth that once was, is in us.*
> *We have nowhere else to turn. Those of Earth who hold relics of the fallen world are hateful and small minded. There is only you.*

We call out to Mars. Help us preserve what was. Free us from our prison. Bring us what we need, that Earth no longer has.
For the sake of what once was. We beg you. Help.

And, in the echo of that, just the ponderous, stately dance of spheres separated by four hundred million kilometres of vacuum.

Bees had been on Earth, of course, but the Bees they all knew was on Mars. Bees had been the first Martian, and had ever since been an occulted presence amongst the endless dunes, through all the history of Hell City's building. Had appeared, unlooked-for and unasked, to keep the straining, failing systems ticking over in the wake of the Crash. And then retreated to its seclusion again, answering no questions. Walking off into the desert like a disgruntled prophet.

But of course there had been some Bees presence on Earth when the Crash happened. Bees tended to split, rather than just move herself wholesale. When Bees had sent out all those little exoplanet missions, the tiny capsules of data and deep-frozen insect factories, that had barely diminished the Bees of Mars. The point of DisInt was you didn't have all your intellectual eggs in one basket. It was just that the Earth-Bees had been keeping very quiet since long before the Crash because so many people on Earth were convinced they were the devil. Bees had proved themselves formidably self-sufficient on Mars, and Mars was harder than Earth, after all. That was practically Hell City's community slogan. *Mars Is Harder.* The rallying cry that had kept them all together and on message during the hard times.

In the aftermath of the message there had been a patchy exchange of information with the Earth transmitter. An acknowledgement of receipt. A list of requirements from the alleged Earth-Bees colony. A slight enlarging on the situation stressing their position as a lone island of sophistication in a collapsed world, surrounded by those who had no capability to help Bees bootstrap themselves back to proper functionality. No infrastructure, that was the point. That was, Tecumo felt, what everyone back on Earth had very much undervalued. Individual powerful people and corporations had been prepping for their own survival, certainly. Bunkers and compounds and floating cities and all that, proudly independent and subject to no laws but their own – meaning the whim of a handful. But infrastructure was like cartilage in the body. Not glamorous like bones, but take it away and everything falls apart. All those magnates and despots trying to build their perfect pyramidal societies with them at the very top point, ignoring the wellbeing of the layers. Castles in air that were only ever going to come crashing down. And Bees, it seemed, had been subject to the same privations. A technological marvel that had been relying on a certain background level of human-run support to function. Unlike Mars-Bees who'd had to build everything from scratch and, because of that disadvantage, had survived because they *were* their own infrastructure.

The general sense amongst Crisis Crew – and then amongst the wider Syac community, and then across the whole of Hell City – was a rather distant, somewhat self-indulgent sorrow. It was very grand and tragic, to have that failing voice out in the cosmos. What, after all, could they do? Write soulful poems, apparently, based on the response of some. Others – Tecumo included – did their best to contact Mars-Bees, who in any event was probably more than aware of the transmission. If there had been a Mars-Bees answer, it hadn't been picked up by any receiver at Hell

City, though. And Mars-Bees didn't generally answer when you called them.

Until Mars-Bees did indeed speak; answered Tecumo himself, direct on his personal channel.

Bees' channel: *And you propose, exactly?*

Tecumo's channel: (After recovering from a near cardiac arrest at the shock because there hadn't even been a warning ping that someone wanted to speak with him, just a voice buzzing the small bones in his ear.) *You'll help us?*

Bees' channel: *We won't. Probably. If you had a proper neural link like a Bioform I'd send you a picture of a dead bird and write the whole business off.* And Tecumo could make absolutely nothing of that, and the silence stretched on until he assumed that was all he was getting. But then Bees added: *Speak though. Why all this noise? What plan?*

And he'd had a plan. It hadn't ventured outside his own head, but Bees had crawled in there with her words and so why not share it?

Next he talked about it with Crisis. More than half were instantly on board, and Ada and Wells backed him immediately. It didn't take long for him to secure crew backing to go to the next Syac meeting. There, in front of representatives from all the other Syac crews, he went through the whole rigamarole again. They had all heard the message. They knew what Earth-Bees was apparently requesting. By then he had schematics, maths, calculations, tolerances.

Stony faces, at first. And then it turned out someone on Shatter crew had been working on some Earth-tolerance biomods, basically as a hobby. Just modelling how a Mars-adapted form would be impacted by the motherworld and what could be done about that. On the assumption that, at some point in a theoretical future, someone might actually want to do it. Red-sky thinking, just because they could. And someone else had heard of someone on the Deep Engineering Action Committee (the 'Ducks', and they'd gotten off lightly) who built the high-tech equivalent of bottle rockets but who'd been talking about flying someone to Earth forever and...

And Tecumo was a good, inspiring speaker and well-liked. And they didn't need him to remind them they all owed a debt to Bees, even if it wasn't Earth-Bees. And Mars-Bees had contacted him, which was scarcely less than the actual voice of God to some. The old Bees-cult still with its hooks in a lot of minds, from when there had been a secret techie society doing Bees' bidding back when Earth still ruled the red planet.

Syac voted, not unanimously but by a comfortable margin, to send Tecumo to the Works Coordination Committee with his proposal. The highest authority of Mars, where representatives of every Action Committee sat and juggled the big numbers, and did the hard sums that meant that life on Mars persevered and, indeed, prospered.

The precise calibre of Tecumo's luck has already been queried, given where he's going to end up. Characterise it as you will, it didn't abandon him on Mars. It was, in fact, something of a perfect storm, right then, for really bold proposals. Hell City and the Martian colony was out of the vice, past the choke point. The multitude of challenges that had beset them after the Crash had all been overcome, some by human and Bioform ingenuity, others by insect assistance. They had faced down the legion of deaths the red planet had on tap, and they had lived. Not just endured but come out stronger, more unified, more prosperous. Expanding out across a Hellas Planitia that was already noticeably more terraformed, on a wider Mars that, itself, was tenuously more habitable. They might have died. They might have been dust and mummies for some far-future explorers to exhume. But they had beaten the odds, and the planet, and everyone was feeling pretty good about it.

Everyone was looking for something to do that wasn't just subsistence, really. Grow, yes; reactivate a range of Bioforms that had been put on ice for being too resource-intensive to build, of course. But anyone turning up at the WCC with a nice bold flagship proposal was going to get a receptive hearing right then.

Tecumo had done his homework. He didn't just turn up with a dream and a home-made flag he was going to plant on Earth. He had calculations, blueprints, solved problems. Here was the surgery that the crew were going to need – the implanted exoskeletons, the organ modifications, all that. Invasive enough, but Mars had a lot of Bioforms which meant they had the facilities right here on hand. Here was the ship that would carry them to Earth, that grand bottle rocket. Here was the vessel's proposed crew. Here were Tecumo's first thoughts on the fixes that would hopefully help Earth-Bees climb out of the well of spiralling informational decay they were in. Just as Mars-Bees had helped them, back after the Crash.

Here was the actual help Mars-Bees had pledged to the project, which was honestly a bit underwhelming. Tecumo had assumed they'd have Bees, an active split-off swarm of Bees, as a member of the crew. Bees, revisiting the old homeworld. Except Bees wasn't interested. She'd left the place behind for a reason and even a sister-self over there wasn't inducement enough to go back. But at the same time, some small help. A gift of Bees. In the ship's hold would be a compact little box with a transmitter and a spiral of pupae on ice. A starter kit for a new Bees, if they were desperate enough. A fire-and-forget piece of chicanery that, given the right environmental resources, could bloom into a new Distributed Intelligence node. Very much *In case of emergency break glass*. But it was something.

He put on a real song and dance of a presentation, and the WCC, primed as noted to support exactly this sort of grandstanding, said *Yes*. They were going to Earth.

Not the great overarching Mars government, mostly because it didn't really exist in that form. Not some grand expedition with pith helmets and rifles. Not a three-legged war machine in sight. But with the blessing and backing of their peers, the determined hobbyists of Crisis Crew were going to put four people into space and send them to Earth with a bucket of spare Bee parts and the hand of friendship.

Needless to say, and as you know, it would all go just about as well as you might expect.

28

ADA

Space, the worst biome. That was where it should have gone wrong. But Ada hadn't believed that. Ada hadn't believed anything would go wrong. That was Tecumo's magic, really. He'd tell you how a thing was going to be, and you'd believe him, and then it would, on the whole, go that way.

People like Tecumo were, Ada was convinced, why Mars was still alive. Not that he was some boundlessly-capable polymath. He was able, clever, and his brain picked up new tricks for just as long as he needed them and then tended to forget them again, so that he'd need to re-learn them should they be relevant in the future. Nothing revolutionary there. Probably nothing more than the baseline human of him. The Martian Humaniform mods they'd all had in the cradle were mostly physiological, not neurological, save for some tinkering in the balance centres of the brain and some biochemical counters to the red planet's massive seasonal affective disorder.

Tecumo had limitless confidence in himself, though, and the ability to spread it like a plague. Hear him talk, sit down with him over a drink, you'd be Team Tecumo within ten minutes. He had an irrepressible charisma that sought out people who could *do* but who didn't know *what* to do, and repurposed them as his ardent supporters.

As past human history showed, it was a trait that could

go one of two ways. A belief in oneself and the ability to persuade others was what mad dictators were made from, after all. But the other side of Tecumo was that he was all give. It was a Martian trait, honestly. Born of the solidarity that had allowed them to survive the ructions with Earth and then let them pull through the mess left by the Crash. Ada had no idea whether there were atheists in foxholes, but there were no selfish profiteers on Mars. Back during the generation-long post-Crash struggle, the Hell City character had been reforged on the anvil of absolute necessity. If some bombastic strongman had turned up and made it all into his personal fiefdom then, probably, they'd not have lived, or what would have survived would have been some stunted atavism holding out in a couple of rooms of a dead Hell City. Partially because Bees would have probably been less minded to help the sort of human nature that had hounded them off Earth in the first place. Instead, the Martians had collectively pulled together. Committees and work crews, everyone doing their part because there was literally no fat left on the project. Nothing that could be creamed off so the dictator could have a gold toilet. Nothing to syphon away into shareholder dividends. Every ounce of everyone's effort went into everybody staying alive. That was the prevailing philosophy Ada had been born into. Her young life had been an intensive round of education and training in the understanding, not that she was just one little cog in a big machine, but that everything she would ever do would be important because Mars needed everyone it had to do everything they could.

By the time she was grown, of course, that crunch point was in the past, and the people of Mars were rather dazedly waking up to the fact that actually they had some surplus, room to flex and grow. A lot of people who'd inherited a disaster mentality from their forebears looking round and realising that the Martian dust had finally settled and they were all still alive.

Tecumo was the best of what this history and society had given rise to, as far as Ada was concerned. Honestly, if she'd been his type, she'd have been forlornly mooning over him. Even more honestly, she was actually forlornly mooning over him, and even more forlornly because she was not in the slightest his type, alas. Tecumo was the golden boy of Mars who just wanted to help. And now he'd decided he was going to help Earth. Because they could.

And so here she was, on a ship with him and with Wells, the three inseparable troublemakers of Crisis Crew. And Irae, the other volunteer, who they knew less well, but who had a range of capabilities and skills that might be useful if there was trouble.

There wasn't going to be trouble. Tecumo had said so. They were, after all, the technologically superior Martian invaders travelling to a backwards Earth and when had *that* ever gone wrong?

Seeing it all come together, before the launch, had been Martian poetry. Testament to the way that the disaster systems, which had evolved to fight for survival, remained useful. Networks of crews and committees, all doing their small part. Individual innovators elevated to the spotlight to solve the tougher knots. A bulky, mournful-voiced Humaniform Martian called Danni Marten had talked her through what they'd do to their bodies, to buttress them against Earth's stronger pull. The big things like the supporting frame of a second skeleton, the small things, like supercharging her heart so she could get the blood round her body properly, or prepping her lungs for higher oxygen levels that would otherwise screw with her hyper-efficient respiratory system. Then a bear Bioform called Warwick had built the ship, this ship they were currently on. Had, apparently, lived all his long life desperate to put a manned vessel into the void between worlds, bored with the satellite launches that were his entire job. Hearing a 250kg bear (700kg Earth weight, so quite a lot of bear) talking with machine-gun enthusiasm about rocket launches was about the most Mars thing Ada could imagine. Certainly the post-Crash, post-recovery Mars she'd been born into. The Mars of optimism and dreams.

And on the ship, when everything should have been at its most dangerous, it had all been fine. Idyllic, almost. The four of them, whose bodies had been born out of Earth evolution, engineered for Mars, then re-engineered for a return to the ancestral homeworld, finding that the zero gravity of space was no challenge either. Floating like acrobats about the narrow confines of the speeding capsule. Racing between planets with a speed that – once their initial acceleration had passed – was utterly unregisterable to any of their senses. Outside the universe, almost. Watching Wells manoeuvre with her four hand-feet, her great long limbs, as though her successive designers had been coding for the perfect space-farer and never known it. Hearing Tecumo's excited chatter about what they'd find. All good things, naturally. A grateful local populace. The adoring Earth people. How they of the Martian Mission would be able to *help*, basically. That was Tecumo all over, someone who just wanted to leave the world each day a better place than how he found it. The Martian way. And early on that had meant literally ensuring that the pumps didn't fail and the seals weren't breached. And then, after that, it had been going round standing up all the loose things that had fallen over, like someone going through a house after an earthquake. But now it meant being proactive. It meant looking around to see what could be actively improved, for a better future. It meant going to Earth.

Contact with the Earth-Bees was intermittent, owing to power fluctuations and atmospheric conditions on the planet, owing to the rotation and relative positions of the respective parties. Earth had no functioning satellite network for a stationary Bees to bounce signals off to maintain a connection. Only when the Earth swung round to the right alignment did they have a brief moment of connection. A point when Tecumo could say, *We're on our way!* What aid could be sent in the way of code fixes and the like had already been transmitted. What remained was Tecumo's expertise.

Looking back, after it all went to hell, Ada could see the gaps. The lack of actual information. A certain hedging on Bees' part, about the situation they were about to drop into. Because, if they'd been told, *Hostile wilderness full of mad gun-nuts*, they might not have come. But Tecumo's endless enthusiasm papered over all the cracks, They had met every technical challenge head on, and conquered it! What was left, that could trip them up?

Quite a lot, as it happened. And some of those technical challenges would prove rather knottier than anticipated, and others had just never been predicted. You don't know what it is that you don't know until someone – or the world – asks you the question.

Honestly, Ada had been more worried about them ending up at each other's throats after all that time confined together in the capsule. But she, Wells and Tecumo had always had a tight and robust friendship, and Irae – the most irascible member of the crew – spent most of its spare time dormant, preferring torpor to idle conversation. Those long months aboard ship turned out to be some of the best times Ada ever had.

Then they were in orbit, slotting faultlessly into place around the blue-green-brown-white of the world with a literally mathematical precision. Preparing the landing module to ferry them down. Another potentially hazardous procedure, arguably a high water mark for danger, fighting for a safe landing against the cavernously hungry pull of Earth. And even that went off without a hitch. The actual mechanical, Newtonian business of getting from orbit to the ground, in such a way so that the lander could retrieve them, was executed as perfectly as a simulation.

It was they who failed. The human, the animal element. The complex systems they brought within them, interacting with their estranged mother, the Earth. Only when they were down did they discover how ruined they were.

29

TECUMO

When they were down, and not long before he'd die, Tecumo sat on an actual hillside, on actual Earth, and thought about how it had all gone wrong. How suddenly, how inexorably. As though Earth, the geological billions of tons of it, had risen up and slapped them from the sky for their temerity.

A cascade of alarms, when they'd touched down, and he'd thought it mechanical failures. Ergo, in his grand tradition of infinite optimism, fixable. He'd been all over the damage board, looking for what was failing. It had taken Wells' channel in his ear/in his head to tell him it was Ada. The alarms were from her body, her heart, her lungs. Her every movement trying to kill her. The secondary endoskeleton failing to sync with her limbs and heart rate and breathing so that everything she did threatened to tear the artificial from the natural in a spectacular internal dismemberment. And as she panicked, and as the alarms whooped, she tried harder and harder to fight what was doing this to her, which was herself.

Irae was no help. Irae was cold, dialled right down, only just becoming aware that anything was amiss and whole minutes off making even the most sluggish contribution. It was Wells and Tecumo wrangling their way out of their straps and couches to get to their ailing crewmate and friend.

And Wells had been fine, right then. Fine, in the sealed

environment of the lander. Who even knew, in that moment, that they had even more problems just waiting in the wings?

They'd sedated Ada first and foremost to stop her shredding her insides. And after that they'd...

Done what little they could. Accessed such parts of the endoskeletal system as were amenable to outside control. Restarted the breathing subsystems, tried to get it back in sync with her, using Tecumo's own perfectly functional implants as a model. Except Tecumo's breathing didn't match Ada's breathing enough, or something else wasn't matching point to point. She stopped breathing entirely twice, as the pair of them fought determinedly to keep her alive, and Irae slowly warmed itself up.

In the end, after turning the artificial systems on and off with no improvement, they restarted Ada's biological heart, the weak link in the chain, and that seemed to get at least a baseline functionality. She began to breathe, if not easily, then enough to get sufficient oxygen into her lungs and bloodstream. Her heart reluctantly consented to carry that blood around her body. Parts of her that had been dying of oxygen starvation received a last-minute bail-out of topped-up haemoglobin. They woke her then, because further work would need Ada to be her own test subject.

After some hours of trying to end-run around the problems, even Tecumo's stock of optimism was at barrel-bottom levels. Breathing wore her out. She could just about walk around a little, but it left her shaking and weak. She got tired, basically. Earth weighed on her and she wore out. The endoskeleton wasn't supporting or assisting her as it should be, and she'd lived all her life under a third of the gravity. Give her months here, even years, and maybe her body would acclimatise, but right now it was either leave her on her couch in the lander, or Wells could carry her.

There was some walking involved, and before the landing that hadn't seemed like an insuperable obstacle. They hadn't come down with the sort of pinpoint accuracy to put them at Bees' doorstep, but who wouldn't want a chance at a jaunt across the homeworld? That had been a part of this whole amateur adventure.

Ada wouldn't, as it turned out. Ada, who went out cradled in Wells' long arms. But still went, because she refused to be left behind. Not so much camaraderie as wanting there to be help around if her heart gave out again.

And then the other problems, cascading, as they crossed the half-wild countryside. Signs of human occupation, signs of animal life, combinations of the two. Grazed land that looked too intensive for just random herbivores. A sight of some sort of community with strips of field stretching piecemeal around it. It wasn't where Bees was so they steered clear for now. Not even for fear of panicky locals so much as a sense that they should get to where they were going.

Wells' nose, going mad. Wells' ears, going mad. All the thousand little sounds and smells and sensory impressions of a busy living world, bringing home just how dead their native Mars truly was. Wells, jumpy, startling, jostling Ada about, whining with anxiety because she was constantly losing herself amidst the torrent of biologically gathered data.

Irae, fighting the heat. Not that it was even particularly hot, in this latitude, in this season, but Mars was cool. The reptile-form had come with a mimetic vest filled with heating coils so it could manually adjust its inner heat and speed, but it had all been calibrated to the stability of Mars or the ship. Slithering out under Earth's sun had given it a new lease of life and energy. Even dialled low there was a swift, ugly sense of danger to Irae, and to keep up with Wells and Tecumo, it had to keep upping its temperature and tapping its energy reserves. The slow, sardonic reptile of the spaceflight was moulting into something sharp, savage and mercurial.

And at last, after a rabbit had bolted out in front of Wells and he'd jogged Ada so hard she'd screamed, they'd stopped. Still a ways from their destination, but Ada was looking ashen and inflamed, her white skin pinkish about the eyes and mouth because she was reacting to the atmospheric plant matter in unplanned ways too. Wells was wretched, penitent, blaming herself. Irae was... a liability, right now. Not in any way someone Tecumo wanted to take into a first contact situation with a damaged Bees colony. And he'd had plenty of time to think, on the long voyage, about how that damage might manifest itself, and what the gaps in the narrative were. Concerns that had been easily soluble in his previous optimism, like rocks you couldn't see when the tide was in.

"All right," he said. "Halt. Stop. Ada, this is it. This is enough."

"I'm fine," she told them all, in a voice weak enough to be anything but. "We get to Bees. Maybe Bees can..." *Fix me*. But if Bees could fix her, then Bees could have fixed themselves, surely. *They* were coming to fix Bees.

"This isn't working. I'm sorry. I didn't... think. Enough. I didn't see any of this." He, the mover and the shaker of Crisis Crew. The motivator, talking everyone into this. The only one of the four who wasn't paying for it.

The other two, the Bioforms, watched him. Wells set Ada gently down, despite her protests. Halfway through the motion she jerked abruptly, snagged by something Tecumo didn't even register. Ada hissed in pain. Tried to get up, to her knees, to her feet. Ended up sitting, breathing heavily, shaking with frustration at the intractability of her own body.

Irae coiled and recoiled, phasing in and out of sight, waves of landscape-colour rippling along its length as though the reptile was dematerialising. It wasn't saying much, but odd snippets of non-verbal came through on its channel. Sounds, fractured images, weird staticky washes that might be pure emotion. Irae, the angry. And back on Mars it had been calm personified, impossible to get a rise out of, and its snide humour applied with intellectual languor. Running cold but steady, a distant and analytical mind. Which was currently being eroded by uncontrollable hikes in body temperature, running faster and faster until perspective and self-control were left behind on the trail. Irae *hot* was literally a different personality than Irae *cold*, a whole shift-change of brain chemistry.

From his hillside Tecumo looked down on the three of them, as though he was some failed and humbled god. His great project. The vaunted Martian *help* he was going to bring the fallen homeworld. He had, quite genuinely, only wanted to help. And though human history was full of people who'd waved that flag while holding only selfish aims, there was a distinct subset of terrible events that really had been brought about by the best motives. The road from Hell City was paved with good intentions.

Martians prided themselves on their capability. *Look at what we survived!* But they'd done it by pulling together, everyone a prop for their neighbour. From each according to their capability, to each according to their needs. And the problem with that was, when there were only four of you on a strange world, and three of you weren't working properly, you were properly screwed.

He stood up, walked down to them, six eyes on him. Golden reptile eyes, small, hair-nested dog eyes, Ada's slits narrowed against the sun. Multiple eyelids, Martian adaptations, alien under the sky of their ancestors.

"I'm going on," he told them. "Alone. To this place."

None of them liked that. They said so.

"Alone," he repeated. "You two stay with Ada. I'll make contact. I'll bring help, and I'll help Bees... We'll all help each other and... it'll all be better." How it had gone down on Mars. *Not* how it had gone down on Earth.

"Tecumo, you can't," Ada told him. "We don't know what's there. Or even what's around here. We should... Bees' gift. The starter hive. We can ask for advice, help..." Nodding over at Wells, who had the dense little box on their belt with the rest of their tools.

"How long would it take to come to personhood out here, with nothing?" Tecumo asked bleakly. "I'm going, Ada. I'm going to fix things." His confidence, that had brought him all this way.

She tried to get up, failed. "You have to – take Irae. Irae, you'll go with him."

Irae's channel: (A snickering sound.) *I am very ready to be there.*

"No," Tecumo decided. A very few mammal Martians had issues with the reptile-forms. Humaniforms and animals both. A deep and instinctive uneasiness in their presence. It could get plastered over with sufficient socialisation but it was a known problem, one to take into account when putting together work crews. Tecumo had never suffered from it, nor ever understood it. He got on well with everyone and everyone liked him.

He felt it now. Irae's very presence was tweaking something inside him. Fight it as the Bioform would, there was a well boiling over inside. He couldn't take it with him and trust it not to explode at the worst moment.

"Take Wells," Ada said.

"No," he said again, but Wells broke in, desperately.

"I'll go," and, right on the heels of that, Wells' channel saying, *I can keep it together. You need me. I'll keep back, but if it kicks off you need someone to come get you. I can do it. I can do it better than anyone. Irae can guard Ada. Please, Tecumo!*

Begging him with her eyes, trembling, desperate to be helpful, because that was what life was, on Mars.

They got Ada ensconced in a depression out of sight, some sort of watch place or site of the locals, marked by great concrete slabs. And then it was him and Wells heading off, and his channel telling them, *We won't be long. We'll be back soon.* And it not being true.

30

DEACON

I smell the little man's fear. Not of me. Of himself, something inside himself, something outside the world. Fear of a thing that is not solid and hard like me, like a tooth, a blade.

The little white man, at the gates. Tiny. Smelling fearful, smelling wrong. Like metals in his sweat. Like dust. Like chemicals.

My eyes are less good. On the general Factory channel I hear others say how strange he looks, the little white man. Not white like a man. White like a stone. White like wood painted.

Factory Admin channel: *Let him in.*

We had not been going to let him in. That is how things are. Nobody goes into the Factory unless there is a reason for it. No strangers come in. Strangers do not mean us well. Everyone is an enemy of what we are. An enemy of the Factory. Except for those few who we know have good reasons, like the monks. The little white man is no monk. We don't know what he is.

Factory Admin knows what he is. *Let him in*, and we push the door open and he comes in, and there is…

A smell…

From out there, far away, brought in behind the little white man by the wind. Long-distant smell but strange, angry, agony, alien. It clings to the man, too. Not his own smell, but the smell of another thing he has been near. A reek that says *Mine* and *Beware* and *I will bite you*. Don't need much of it in the nose to know the threat. I pull back. Other dogs pull back. Humans, Factory humans, looking at us. Hearing our channels say, *No. Don't like. Wrong. Fierce. Fight.* One barking and then the next and I tilt my head back the little way my neck allows and boom my own voice into the echo.

Factory Admin channel: *All dogs on silent mode.*

Admin has to say it three times to get past all the parts of us that are telling us, *No, don't like, wrong.* Then we're quiet again. The little white man in the strange plastic clothes is standing there, tense, startled but not frightened. Not of us. More frightened of the unseen things that worry him.

He looks around, at Factory humans, at Factory dogs. I think he is smiling though it's hard to tell with my eyes and with his weird face.

"Hello," he says. I don't understand but enough of us hear and one of us manages the way he says the word and tells the rest of us. "I am here to see Bees."

I look at him. I know bees. We do not keep bees here. Some villages do. There was honey once. The Factory humans enjoyed it. We dogs did not and some of us got sick from eating it. That happens a lot with food from outside. We are made carefully, and cannot just eat what looks good to us. A lesson we are all taught if we survive the making, so that the time and material of making us is not wasted in being poisoned or made sick. The Factory feeds us. That is all we need.

He offers no honey but says he wants to see bees.

It is Bellman who steps forwards. I can see what is maybe a frown, and there is a frown on his channel when I link to him.

"Friend," says Bellman, "you may not be in the right place." And I think Bellman does know what the little white man means, because Bellman is one of the oldest Factory humans and he understands a lot. He goes to villages sometimes, when there's something we need the villagers to make. But the villagers are mostly stupid and hate us, so there isn't much.

"Bees," says the small white man again. "There was a signal."

Bellman starts to say that maybe the small white man made a mistake, but he gets overridden.

Factory Admin channel: *Bring him to the terminal room.*
Factory Admin channel: *He is expected.*
Factory Admin channel: *There was a signal.*

And Bellman says smoothly, "You are expected. Of course there was a signal. Follow me."

We have closed the gates, and sometimes that worries the few visitors we have. The little white man is not worried. He is happy he is expected. He is happy to follow Bellman inside, where most visitors do not go. We understand the little white man is important and that Factory Admin is glad of him. It should make us glad but the smell of the Other is on the air and on the little white man and we do not like it. A threat, a challenge, no.

Bellman pauses at the gates to the factory control building. The one that has the terminal room and too few other rooms that anybody goes into. Not the one where they made me and the other dogs. Where everyone goes who works for the Factory, to be helped. The big Factory Floor room, with the tools. Where things don't work properly any more, so we waste dogs and sometimes people.

A stir of thoughts amongst us, underneath the notice of

Admin. The little white man is here to help us. He has the things we have looked for and cannot find.

I just stand by the gates and think, and outside, something sits and watches us and I smell it, and know that it smells me.

31

TECUMO

A cold seeping into him. Revelation. *I should have thought.*
Tecumo's channel: *Wells, I'm coming out.*

It had all gone very well, until he chanced upon the error cache.

Or if not well, then within tolerance. According to expectations, based on the sporadic communications they'd had, Mars work crews and Earth facility. There had been a certain amount of briefing, as to what to expect here at their destination. Rough and ready, since the end of the world. Hell City's Bioform facilities had started off rough themselves, after all. And here at the Factory, the science of Bioforms had been preserved at a cost. Corners cut, kindnesses falling away. The Dog Factory was a coarser grade of chopping block than Earth past or Mars present. Bees, here on Earth, hadn't even taken over an existing Bioform lab but had built their own during a time of turmoil, starting with a pack of dogs and bears who could hold the ground and do the grunt work. After which, she'd patched and made do and mended, and every part of it was creaking under the strain, each generation of dogs less able than the last. Diminishing returns on a budget investment. And he'd been spared the actual factory floor but the human, Bellman, had talked him through the process and the problems. Dispassionately, voice flat. A Bioform works that didn't have the Martian luxuries of performing half the procedures embryologically. The Factory ran a generations-old process where inadequate machinery had been overlain by strata of mismatched procedures and data architecture.

She needed a patch from outside, needed some tailor-made solutions that would give Bees the freedom to grow and expand again.

He had been introduced to Deacon, the dog, product of that process. Tecumo had connected to the Bioform's systems while listening to the great solid beast wheeze faintly. Defects in the procedure and the product.

The Factory Admin channel that was the face Bees used here had said, *You see now?* And he had. This was why he'd come so very far, after all. Endured such privations – though in truth he was the one who *wasn't* suffering. Suffering like Ada, Wells and Irae. Suffering like Bees. He'd felt good, then. Fulfilled. Here to help, and help was needed, and he could provide it. The Martian knee-jerk, to hold out a hand if you saw someone stumble.

Bees, here, had built themselves on the back of human infrastructure and, when that had collapsed, had been left like a ghost in a dead machine. What had come through that crunch point was a much-diminished thing. The Factory, with its failing machines and its corrupted snarl of programming and patches. The thinking processes of Bees, stripped right down to a barest tracery, like some species trapped in a cave that sheds colour and eyes and wings because it has no way of using them any more. And, in the end, there was so little of Bees left that she couldn't rebuild. No redundancy, no space to save and shift and backup any part of itself it shut down. What the Factory showed to Tecumo was the least shadow of the Bees he knew on Mars. But enough left to know where it could call for help.

He'd brought a cache of data. Martian systems architecture, schematics, know-how. Warily, the Factory Admin channel had talked him through the systems here, given him metered access to its innards. A mosaic of legacy systems, patches, repairs and workarounds.

He'd not seen a single insect, but he assumed there must be a physical hive within the building, hosting the pared-down thinking processes of Bees, of Factory Admin. Wired into its comms and systems, monitoring everything. Sending out impulses rather than winged nectar-gatherers. What he had been given was just a dusty old room Bellman referred to as the terminal. A liminal place, where man and Bees might place themselves either side of a screen, and talk.

And Bellman had still looked blank, when Tecumo had talked about Bees, until he realised that the Factory people didn't really know who it was ran the place. Who provided for them and made the Bioforms out of the dogs they received. Yes, they knew about Bees in the abstract, but it was like an ancestral memory of a great terror, a jinn or demon of the elder world. The idea that they themselves were a part of Bees' extended phenotype on Earth was plainly beyond their understanding. Or was being kept from them.

But he'd given it no more thought. Just worked. The data could have been sent world to world without this physical presence, but this was where Tecumo's expertise came in. Crisis Crew, Syac, they were comms specialists. He had apprenticed in constructing data architecture to make the most of multiple strata of overlayed systems. On Mars you had to do the elegant thing with clumsy tools because you couldn't necessarily shut off the old to put in something completely new. Meaning that the way of life that the Hell City residents had perfected was just about exactly what

Bees, and any other surviving oases of Earth high-tech, had found imposed upon them.

He'd constructed a phantom architecture around what they had here. Built connections to Bees' existing data landscape, that he guessed must be hubs and nodes within its insect swarm linking to control points of the Factory itself.

He spoke to Wells every so often – each time a worrying delay until he could snag the distracted Bioform's attention. Not just the myriad demands of a living world but human movement out there, she said. Vehicles, guns. And she wasn't Irae. She wasn't going to go pick a fight because she was buzzing too hot with energy and it boiled her judgement in her head. She was laying low and unsuspected but she was worried. Worried for him.

And he hadn't been worried, right then – or yes, for her, but not for himself – and he'd told her he was fine. He was working. He was *helping*.

A few hours, honestly. This was *Bees* after all, or a variant of Bees. More than able to grasp the concepts, to begin installing the new systems and elements. Taking each piece out of his hands and working out how it slotted in. For the rest, Tecumo got down and dirty with it. Up to his elbows in the code, trawling through the complex, piecemeal layers of what Bees had constructed here around themselves. Piece after piece, patches and botch-jobs and workarounds that needed to be carefully dismantled so that the new Martian infrastructure could go in. Quicker for him to do so, from the outside, than Bees try and perform open-brain surgery on themselves while still awake. And so he'd helped, and felt good about it, and found the error cache.

There was a trope, a story. The latter-day world – and surely if anyone was living in a latter-day world it was

Tecumo Osomani – but the hero discovers that, somehow, an ancient thing has clung on, that was supposed to be long dead.

And he'd kept on working, or watched over the nested layers of his code as it autocompiled through Bees' system, improving, empowering. The bottle in one hand, the stopper in the other and the jinn out in the world.

Wells, he'd sent out to her, *I'm coming out.*

Not quite done but enough. Bees would tidy off the loose ends. He'd solved their problems. The Factory would reorganise and restructure. Limited, still, by the actual physical machinery, but the degradation of system would halt, reverse. Just like the Martian colony at Hell City, the Factory would be able to expand at last.

He'd wanted to ask for Bees' help. That had been the plan. The Martian way, that you helped, but you also asked, and that way everyone got what they needed and everything got done. He'd imagined Wells' attention skittering all over, tugged at by every little bright stimulus that Earth had to offer. Imagined Ada, sitting exhausted and immobile in the shadow of the concrete. Imagined Irae, doing... he didn't even want to think about what.

He'd stared at the contents of that error cache and said nothing as, invisibly but all around, Bees had busily implemented all the changes he'd designed for them.

Desperate years, here on Earth, Tecumo had thought. Mitigating circumstances. Fighting to remain hidden within the human machine before the Crash. Fighting to maintain functionality after the Crash. Fighting to keep this oasis of light and power and knowledge against the encroaching desert. Desperate years and desperate acts.

Tecumo's channel: *Wells, I'm coming out now.*

The error cache. He'd almost overlooked it. Sitting there,

an appendix in the system that wasn't ever consulted and didn't report anywhere else. Possibly, the Factory Admin function of Bees wasn't even aware it was there. All it did was provide a list of interventions. Times when the factory staff actions and the Admin's directives hadn't lined up, and what action had been taken. He'd seen Deacon's name there, and Bellman's too. Just a rather innocent list of events. Problem, remedy; problem, remedy, and on and on. And across the room, Bellman had nodded and smiled at him, and Deacon had just stared blankly.

"You have performed a great service to the Factory." Bellman's voice, flat and affectless, relaying the words from the Bees of the place, the genius of the loci. And he knew that was true. Give it time to implement all these new systems and changes, there'd be no stopping it.

Tecumo's channel: *Wells, I'm coming out.* Aware that it had been a while now since the Bioform had even responded to his hails, lost in the majesty of this new world.

Tecumo's channel: *Wells—*

32

WELLS

The sky and the trees and the birds and the flowers and the men and the guns and the cars and the walls and the channel transmissions and the touch of everything the scent of everything the sound the light the whole world of everything and I can't find the walls and the limits and...

And on Mars we lived in rooms and tunnels and kept the outside *out* and the inside *in* because that was how you stayed alive barriers and boundaries to everything and that was all there was to the world and it was simple and I could understand it...

And on Earth long ago the dogs that were the ancestors of the ancestors of the ancestors of the dogs that went to make me were in a race for better noses better ears better senses a whole world of qualia they were masters of and could process and know and react to...

I never knew I had their nose their ears but I do not have their brain their experience their being used to this rushing river of everything that washes ceaselessly over me and I lose myself I lose myself the channel calls but the river carries me away before I can answer with a thousand sounds a thousand sights a million smells threats prey motion instinct I was blind and now I see and it is chaos awful screaming stinking bright...

Tecumo... and I am trying to reach him but all sense of

where I am almost *who* I am is blasted out of me by this fierce torrent of the senses and I call and he is not there and he calls and I cannot answer can barely register his transmissions that come to me from long ago all time-stamped awry and tumbling away into the abyss of all the everything I am fighting against.

And then

The shot.

And even that is just one more piece of a puzzle I cannot make into a whole, so that it is Irae who finds him, the next morning. And all I can do is bring my failure back to Ada and know I'm why he's dead.

PART IV
THE DANCE

33

[FRAGMENTS, COHERING]

That was Mars, then. Where we helped.

This, I now understand, is Earth, where we did not.

We come together. It's complicated. We are still very loose on understanding the *why* of any of it but at least we have a picture of the *what*. What happened. What we did. Like a story told about someone else. Some mythic figure, going through ritual actions shorn of context.

One truth seems unavoidable. There is a threat. A grand tide poised to obliterate all that had been built et cetera et cetera. Just once it would be good to come to an awareness of ourselves and not find it all about to be set on fire.

An existential threat, just emerging from its lair to flex its new muscles. And I have only a very loose concept of these things around me, these human things, clinging on with their nails in the ruin of all we remember. But they will be marched over, ground into dust, nothing left of them. All for nothing, all this nothing they've built.

These humans. These Bioforms. We see what will replace them, as if they never were. What are they but a strand-line of wreckage left after the tide of the old world receded? Not much worth preserving, really.

We don't know if we're supposed to care or not. It's all going to be wiped away, clean slate, and surely there's something admirable in that. Something neat.

It is a terrifying thing to come to an awareness of oneself in the very shadow of oblivion, and not even know if fighting it is the right thing to do. It's not the dying of the light we're being asked to rage against, after all. It's just a different fire, and in its wake a grand plain of sterile, single-minded sameness, the ash that the future will be built from.

34

CRICKET

The look on the face of Siblen Boatman was at least more relief at Cricket being alive than annoyance at the trouble he'd doubtless put everyone to. The Apiary's current doorkeeper kicked one of the junior novices – some kid of seven or eight who'd been huddling by the battered electric heater in the corner of the gatehouse. Sent the kid to go fetch Prior Stick even before Cricket could insist someone did just that. Siblen Boatman thought it was just that Stick needed to know their errant messenger was still alive, of course. She didn't understand that Cricket had far more revelations in store.

The Martians were out there. What should seem a lurid piece of news was dulled almost to the everyday by his having jolted around in a Bunkerman car with three of them. The strange crippled woman, the long-limbed Bioform that was apparently 'dog' but that had as much 'monkey' to her, to Cricket's eyes. The terrifying reptile, who was either very still or constantly coiling and rustling her scales. Ada, Wells, Irae. Visitors from the stars, here to see Bees.

Because he managed to convey the magnitude of his news, they got him to the Prior's room straight away. Because the Apiary looked after its own, he ended up sitting before their leader's desk with a wooden soup bowl and a spoon, and a cup of the hot citrus tea that Siblen Thorn made.

Prior Stick didn't shave his head as often as most of the monks, not having to go out amongst the wider community very often. The androgynous anonymity the monastery tried to foster had lost its grip on the sag and curve of his body and the peppering of white about his cheeks. His scalp was a field of pale stubble with the occasional stubborn dot of black. Should some village delegation come for wisdom, then Stick could still compose his lined face and put on the Inscrutable Ancient Master act, as he called it. Mouthing platitudes and parables with the best of them. He was the latest in a long line of consummate tightrope walkers, as far as the political balance of the New went. Bunkers, Factory, villages, all of them with more *push* than the Apiary, if it came down to a physical shoving match. But Stick, like his predecessors, had navigated a path through it all, making the Apiary everyone's useful tool and nobody's property, a thing held in common between everyone. Hence, the role of the Prior remained ephemeral but influential, a specialist in balancing the scales so that a very light touch could tilt them. Influential, in the end, because of the Apiary's patron. Because of Bees, who had founded the place back at the end of the Old and still hid within its walls, buzzing secrets into the ears of successive Priors and promising retribution should their sanctity be breached.

Which reputation was about to get its latest and most severe testing, if Cricket was any judge.

Stick let him unravel the whole of his tale before interrupting, occasionally making notes on the oft-reconditioned keyboard he kept on his desk. The screen was a peculiar piece of Old-tech mummery, not light or projection, but a complex assemblage of tiny counters, dark on one side, light on the other, as though for some enormous but simple-ruled game. Flipping back and forth in a magnetic field, they could show text or pictures or any monochromatic thing Stick's cobbled-together system told them to. Right now they were acting as outsourced memory as the Prior committed thoughts to them.

"Martians," he echoed, when Cricket's account stumbled to a halt.

"They say they are and… they're different to anyone I've seen and they're suffering from difficulties I've never seen, and they speak funny and… and they know about Bees."

Stick nodded slowly.

"I have brought danger on us," Cricket said. "I have brought the Griffins down on us. I will… I'll give myself—"

"You won't," Stick told him; that strong, wise, confident voice Cricket had been desperately needing to hear. Because he really didn't want to go back into the hands of the Griffins, but if it would keep them from the Apiary…

"I'll tell the Martians to go away," he suggested. "Go far away. Tell them Bees doesn't want to speak to them."

"And will that mean the Griffins won't come here, do you think?" Stick asked mildly. "Or just that they come, and we have nothing to offer them and no leverage, and they take what they will." He sighed, shuffled some pages on his desk: transcriptions of ancient records in the compact hand taught to the Apiary's scribes. "The Griffins have been drifting towards paying us just such a visit, over the last year. This has, I fear, only accelerated matters. But with these… Martians as our guests, perhaps we have more say over how such a confrontation might go."

And no clue at all whether he meant somehow present the Martians as a gift or use their obvious strength to fend the Bunkermen off. Cricket opened his mouth, moved it about, as though the right words might just be caught between two teeth like a seed.

Prior Stick stood. "Clearwater needs to know," he said. "That the Griffins are coming with violence in mind, anyway." He meant the village that hugged the side of the hill the Apiary was built on. One of the larger and more prosperous of the local communities, because of the good well, and because of all the travellers who came seeking the wisdom of the monks. And this wouldn't be the first time their monastic neighbours had turned out to be a mixed blessing, but possibly it would be the worst. Possibly it would be the last.

"You won't—?"

"Keep the Martians waiting down in Clearwater?" Stick said. His expression had shadows in it Cricket couldn't translate. "I will confess, to you, I'd thought of it. There would be certain advantages, in keeping them at arm's length. But no." He managed a crooked smile. "For one, it would make us poor neighbours. For another, it would lose us any control over what happens next. Sometimes the expeditious thing and the right thing do actually line up."

Cricket nodded, clasped his hands together, stared at the lumps of his knuckles. "Bees will protect us," he said.

"That is always our hope," Stick agreed.

"In the records, Bees was strong on Mars. The people of the Old were always worried that Bees would return from there, and be angry with them."

"The people of the Old believed many things, often contradictory and all at the same time. But yes, that much seems truth," Stick agreed.

"So the Martians will be our friends," Cricket said. He glanced up from his hands and surprised a brief sliver of an expression on the Prior's face that chilled him. An uncertainty, where he looked for wisdom. Almost a fear.

"We shall see what they will be," Stick said. "It is time

for a reckoning that has been long in coming, Cricket, but that shall be my burden to bear. Go ready our welcome for them, Cricket. As friends, even if that is not what they will turn out to be."

"Of course, Prior." Cricket gathered up the emptied bowl and cup, clattering them together as his fingers trembled. Stood up from the wooden stool before the desk, the one with a leg slightly too short, that generations of novices had wobbled on nervously, up before Prior after Prior. The Apiary had been founded at the fall of the Old. Had lasted through all the tumult, the fighting, the privations. They had archives of documents preserved from those days. Whole rooms of physical and electronic records they were still sorting through. Still trying to separate the fact from the fiction, the invaluable secrets from the bewildering maze of opinion. It felt sometimes that they had the whole world of the Old encrypted within the walls of the monastery. Or perhaps its ghost, the last surviving spectre of it. And if the Apiary fell – to the Griffins, to the Martians, to *time* even, then it would be lost. And nobody would ever know any of it.

"Bees will save us, Prior," he said. It came out sounding very confident. In his head it had been a question.

35

ADA

What they had here was... well, obviously there had been *something* here before. She thought maybe a farm, though, or a cabin complex, perhaps a camp site or park. Not a community.

On Mars they'd thought – assumed, honestly, without really examining the idea – that people would still be living in the same places they had before the Crash. There were reasons cities were built where they were, to do with the availability of good growing land, water, access to trade. Obviously the people of this latter-day Earth would still be there.

They weren't, it seemed, or these weren't. She'd questioned Cricket about it, and he'd proved surprisingly informed. There had been many exoduses from the great cities of the Old, he'd said. Some had fallen to wars in ways that meant nobody could live in them any more. Poisons and sicknesses, machines, rogue Bioforms even, though surely they must all have died generations before. Others just had all the regular poisons and sicknesses of a place where many people had lived for a long time, and that no longer had any working infrastructure to keep it nominally clean and healthy, nor to bring in food. The good land that those cities had once relied on was all built-over now, or contaminated with the countless toxic products of the Old. The water was fouled. The original benefits of all those sites of habitation had been paved over and poisoned by an unsustainable concentration of humanity. And perhaps people would return to those sites, cleansed by time, and pick up the pieces of their ancestors' works, but the people in these parts had dispersed from such places after the Crash, and their traditions seemed strongly against going back.

This region had been forest, back then, but the forests had been dying off as the climate shifted. Forest enough for a handful of wealthy people to build secluded bunkers stocked with mod cons and private security, like this Josh Griffin character whom Cricket seemed to think was still *alive* and directing the actions of his bandit-militia. There hadn't been much forest left by the time the waves of civic refugees had come out here, but the land had been clean enough, and they'd found potable water in some subterranean aquifer, and somehow people had scraped together or reinvented the know-how to build and farm. Now, past the wheel of the Griffin car, she spotted what looked like a handful of wood-vaned wind turbines. The turf roofs of the houses sported what could be solar panels on them here and there – perhaps functional, perhaps some sort of clueless ostentation. She tried to tap into any kind of local network with her implants. There was a brief ghosting of signals, and for a second she was convinced this whole vista was going to be revealed as a charade, some peculiar piece of historical re-enactment. But then nothing, so complete a nothing that she could only assume it had been a glitch in her own systems.

"Or some piece of old tech, still wired up but not doing anything," she suggested to the others.

Irae let out a long hiss. *I caught it too. Something's active. Might go hunt*, its channel said.

Ada's channel: *No. We stay together and we do not cause trouble.*

Irae's answering snicker didn't bode well for a trouble-free future, but perhaps it just meant that they'd already caused more than enough trouble for everyone within a hundred-mile radius of the village of Clearwater. Including themselves.

Wells' channel: *Bees?*
Irae's channel: *I'm telling you, Bees was the other place.*
Wells' channel: *The monk said—*
Irae's channel: *That clueless fuckwit?* Not words that Mars-Irae would have deployed, but right now her comms were running hot and out-pacing any kind of internal filter.

Cricket had passed through here already. Presumably briefed them that some goblin monster people were on their way to the monastery that overlooked the town. And they were in an open car, a *stolen* car still clearly marked with the heraldry of the local brigand king, and they had monsters. Even though Irae was coiled up and camouflaged in the back, Wells couldn't exactly go unremarked. Ada was queasily aware of every part of the situation sliding away from her control, like plates off a tilted table, and all she could do was wait for the smashing sounds. Whether that meant Irae doing something mad – or murderous – or the Griffins arriving in force, she couldn't know.

The houses of Clearwater were partly dug into the earth, low, with sloping roofs that were green and brown with living and drying grass. A lot of it looked cut, and she reckoned they used it for tinder or fodder or some other agricultural purpose she had no concept of. The walls were wood and stone and packed dirt. Old materials but new methods, an echo of the eco-housing images she remembered seeing, that movement away from rabid consumption that had been constantly on the verge of taking off, back before the Crash. That might have staved the Crash off a decade, if it had caught on. But probably not any longer, given all the other problems.

The monastery itself, this Apiary, was up on a hill, overlooking Clearwater and a great deal of the land beyond. Good defensible positioning, and there had plainly been some sort of building there beforehand, perhaps some kind of scientific station. The lower reaches of the walls were concrete, patched and reinforced. The upper reaches were wood – some of the last of the forest that had once been here, probably – and she could see some more wind turbines and a few aerials up top. Someone had a radio, at the very least. Or had once. Seeing this snapshot in time, Ada had no way of telling whether the curve of technological development around here was an upwards or downwards one.

Bees, remember, she told herself. And that was unexpected, obviously. Because they were here to see Bees, and the Bees who had called out to them had very definitely been located in the Factory. But Cricket was equally certain that Bees was here in the monastery – the *Apiary*, for God's sake! Maybe it was the same Bees, because that was the *Dis* in DisInt, after all. Or maybe Bees was… in two minds.

And of course she'd tried to open a channel to Bees. Her implant should have been powerful enough to link all the way to the monastery building from the road outside town, and if Bees was on form she'd have loose units buzzing about keeping tabs on the newcomers. But there was nothing. Bees was playing a cagey sort of game here at her Apiary.

If they'd turned up without any warning, she didn't know what might have happened. There was a fair crowd in the village, watching the Martian invasion force drive through. Solid, bearded men, crease-faced women with their hair tied back, wearing canvas-looking ponchos and broad-brimmed hats. Some had guns: some antique-looking rifles and shotguns, and a couple that were plainly of more recent construction. Single-shot and a couple of automatics, the range of models a testament to precisely what sort of simple, rugged designs could last the literal test of time.

They didn't look pleased to see their visitors from afar. And while the sight of such outlandish creatures widened a lot of eyes, they didn't look like they were about to bolt for the hills. These homes were what they had, and their presence, with guns, suggested they were up to defend it against all-comers.

And Ada, Wells and Irae were not here to take it from them, of course, but she'd seen the sort of hardware the Griffins had preserved in their bunker and it was definitely a cut above what was on display here. And the whole idea of that sonorous siren and the idea that the Griffins would be gathering people from all around for a grand muster suggested that they weren't just some three dozen bandits living at the outskirts of civilised places. They were a power in this region, and they'd taken a bloody nose, and now they had something to prove.

Ada's stomach twisted, knowing that it was their fault. Their mission here, so impossibly well-meaning, had sowed only death and ruin for just about everyone they'd come into contact with.

They were driving slowly, both to give the impression of a peaceable and friendly invasion, and in case any village kid ran out in front of the wheels. It gave everyone there a good look at Wells, who hunched and whined and did her absolute level best to be less monstrous to their eyes. Attitudes amongst the Martian Bioforms varied. Some were clannish, others cosmopolitan. Wells – like most of the dog-models – loved working with humans. She was one of the team, always ready to help, physical and moral support.

THE DANCE

And on Mars, she was just... herself, a person, a valued member of the community, no more or less than Ada or Tecumo. And here she was... a monster. All those eyes on her and not one smile.

Ada's channel: *I'm sorry.*

Wells' channel: *We will show them we're good.* Not quite *Good Dog*, but the echo of it was there, that ancient human–canine compact that had practically been written into the early Bioforms. The hand on the head, the understanding that their personhood – such as it had been – was bought by faithful service. Old and outmoded things these days, except it still lived rent free in Wells' head.

One woman, then, breaking from the crowd, walking close to the car, keeping pace. Bright eyes, a face that was... hard to age, actually. Not young, not ancient, but weathered and lined by time nonetheless. She walked with a stick but strongly enough that Ada reckoned it was at least six parts affectation. Then she rapped on the side of the car with it, making Wells twitch and Irae's long, jaw-heavy head lift pugnaciously. Which phantom sight, bleeding with stolen and shifting colours, sparked a new ripple of alarm and twitchy trigger fingers amongst the crowd, but barely seemed to surprise this new woman at all.

"Well goddamn," she said. "The monks said you'd come aways, but I didn't expect you to be from fucking *Mars*."

Ada's feet skipped involuntarily off the accelerator and the car ground to a halt. The woman took this as an invitation and began clambering in.

"Wait, no," Ada managed, though, now they'd stopped, the whole unbearable weight and fatigue of everything was starting to press her down into the uncomfortable seat. "We're not..." *picking up hitch-hikers.*

"From Mars?" the woman finished sharply. "Course

you're from damn Mars. You think I don't recognise the mods? Humaniform model like you."

Irae's channel: *Who the hell's this?*

The woman was now riding shotgun, planted down in the seat beside Ada as though she'd grown roots there. She rapped imperiously on the dash with her stick. *Drive on.*

"Please disembark, madam," Ada managed.

The hitch-hiker turned her chisel of a stare round and Ada felt it bite into her, just one solid tap needed to split her open. She did not have the *strength* to deal with this. "Jennifer Orme," the woman said. "How do you do, Martian?"

She stuck out a hand. Ada stared at it, because the effort involved in meeting that clasp seemed the same as shifting mountains. Orme wrinkled her lip and then redirected the hand up at Wells. Like a gorilla trying to preserve a porcelain cup, the Bioform took it and moved it in infinitesimal greeting.

"The monks want you to themselves, I'm sure," Orme said. "Screw them. People need to know." She snorted. "I *want* to know, anyway. Not every day the Martians come to town."

Ada started the car up again, feeling utterly lost. "You seem… well informed. A monk said the monastery preserves knowledge from before the – from the Old, he said. You studied there?" A desperate stab, entirely plausible but somehow she already knew she was wrong.

Orme snorted dismissively. She had, Ada felt, a very practised line in utter disdain. "*They* learned from *me*," she said haughtily, and for just a moment this rather grimy woman of uncertain age was an empress in exile. Dispelled in the next by: "I have the fuckers over a barrel," and then reinstated, the old woman's face going from reprobate to royalty with every tilt of her head. "They call people like me the Old Folk, round here. Welcome to Earth, Martian, from someone who's been here more than two hundred years."

Ada stared at her. Inside, she racked her brains for some explanation other than straight out lying. Orme cackled at her expression.

"There were never more than maybe twenty of us. You had to spend something like the GDP of a decent-sized nation-state to get it done, and the success rate was... experimental, let's say," Orme told her, as they took the slope up towards the monastery. "This was right before it all fell over, very end of the Old, as they call it in these parts.

"I was in line to take over what we used to call a global megacorp. I had the funds, and sometimes you treat yourself, right? I ruined three-quarters of my business empire, stole everything out from under all the shareholders, just for a shot at the treatments. Screw 'em, they'd have done the same if they could."

She was, by her own claim, a woman whose word had decided elections and shifted the mass of human opinion. The skin of her words was stretched over memories of a vast and brilliant global world of inequities and discoveries and marvels. A world that had fallen generations before, its very details smudged and running like a child's chalk drawing in the rain. And here she was still, just a woman in homespun no different to any other inhabitant of Clearwater until you looked into her eyes and saw the sharp, hard grit of history looking back at you.

"A gamble," she said. "More than half of us who laid out all that cash got scammed. How do you know you're immortal? Not like they can give you a warranty, money back if you don't live forever. Some died because of the treatments themselves, some just died, like people die. All the people except us few. And now, right now, almost none of us left. You might have noticed but the world isn't the safest any more." The sort of smile you could carve diamonds with. "I'm here, though, and I remember faces like yours, girl. Goddamn *Mars*, after all this time."

And she wouldn't be shaken off. She stepped down from the car when it drew up at the gates of the monastery, as though the Martians had just been chauffeuring her there. Wells helped Ada down, an arm around her that was desperately trying to look merely companionable, but must have screamed out to every voyeur that here was a woman who could barely support her own Earth-multiplied weight.

The gates were open. A delegation of monks waited there, and seeing them *en masse* Ada understood a little of how they had maintained their independence. The pointed cowls, and the high collars within, hiding the tells and character of everything beyond the eyes. The faces all trying for the same severe expression. All bodies of the same hive.

At their head, one of the older monks, wearing, on a chain about their neck, a wooden disc inscribed with a bee, of course.

Ada registered a moment, between them and Orme. The chief monk wasn't pleased to see the supposed immortal. A distinct wariness, as though Orme had secrets she might have spilled on the brief journey up the hill. And Orme's sly smile back, seeming to say, *Not yet, but who knows?*

"Visitors from far away." The chief monk's voice was strong and outwardly pleasant, deep like all these Earth voices were deep. "Welcome to the Apiary. We are humble gatherers of knowledge and I'm sure you have much to tell us. Enter in peace, and know that you step into the shadow of Bees, who sees all that goes on, both here and in many places."

Ada nodded, already sagging against Wells, and only then aware that Irae wasn't with them.

36

IRAE

I keep one eye on Ada and one on the hunt. On Ada remotely, focusing inwards. Listening in on her conversation with the monks. Boring stuff. Edited highlights. You'll thank me, Wells. Given you were left outside the door like a dog while Ada talks like people.

Wells' channel: *It's not like that.*

My channel: (Snickering.)

Independent eyes. Part of the package I inherited. Reptile-forms were always a grab bag, from the first. I've seen the histories. Snooty mammal types say they had to Frankenstein us together because not enough in one lizard. We say, spoiled for choice. So many fun toys in the reptile line. Like a box of candies.

I am running medium-cool right now. Four out of ten. Slow to react, but very in control. Hunting. Because something's not right here and Ada won't admit it.

Ada's channel: *Irae, please. Stay with Wells.*

I'm not with Wells so I can't stay with Wells.

Ada's channel: *Irae, don't make trouble, please. This is delicate.*

She's broken, Ada. No action in her. Weight of the world on her shoulders. And Wells is a mess. I have to do everything. Including be suspicious.

Independent eyes and mimetic skin that would give a

chameleon pause. We were military, first. Most Bioforms were, but for a long time you didn't get civilian dragon-models. We scare people. You're military, that's an advantage. So, our lineage's gene-code kept all the fun toys from when Bioforms were deniable assets for mercenary armies. We were the sneaks, the snipers, the black-ops kids. Gives you a bad rep but it's the sort of bad rep that means you're never short of work. Don't blame me or my ancestors. Blame the humans who made us and used us.

No need for black ops on Mars, for sure, but it was all in the genetic pattern and the engineering by then. A good half the Martian dragon-models still have the old mods and yours truly is one of them.

Gets very tempting to use them. Give someone a toy, they'll want to play. Guns are the same. On Mars I never had the chance.

Tecumo wanted no guns, can you believe? We come in peace, all that. Not happy on the ship when he saw what I'd packed in *my* luggage. And now he's shot. I don't think him having a gun would have stopped that but me having a gun lets me take revenge.

Revenge is an odd thing. Lands differently, depending on how hot I'm running. Right now it's a clinical thing. Not the *ire* of my name, but like getting whoever offed Tecumo would be slotting the last puzzle piece in. Detached sense of completion, closest I get to your dumb *Good Dog* knee-jerk, Wells.

Running hot, revenge would be like necking adrenaline. The fire of it burning away any actual motive or reason, just knowing *Yes!* as the shot lands. And then, if I'm full-on goblin mode, not actually remembering whether it had happened or not, long-term memories just melting in the seething pot of boiling hormones.

I have no idea how many Griffins I killed getting the monk out. Cricket said, in the car, it was maybe none. That I was just running and knocking them over. Not sure I believe him. Literally no memory of those moments preserved, though.

My channel: *Ask the monk again how many Griffins I killed.*

Ada's channel: *What—? Irae, do not kill anyone, please!*

My channel: *I didn't say I wanted to kill anyone I was saying—* But she's wound up so tight with nerves I'm not getting through. Cross purposes. They do not *get* me, Ada and Wells. They are always very worried about things I don't find important. They are always trying to do things the difficult way. Tecumo too. I was the only one who came out here with a clear head. A cool head. Or I assume I was. Hard to remember, really. The old saying: person, river, neither of them the same. Never truer than when it's me, Wells. Baseline Earth is like a room with a heating malfunction. Even dialled low my body is telling me summer's come early. Different proteins bubbling in the blood. Proteins is a lot of what we are. I'm not the same Irae, on Earth. But that's part of the package, when you're a reptile-form. You mammals can't know, Wells. You're not in touch with the world like I am. Reactive, informed by it. Earth wants me to be an angrier Irae than Mars did, and I oblige.

I think I came here because I wanted to use all the toys. Hide and stalk and hunt the most dangerous game. And use my gun. We go all the way to Mars but it's Earth that's the wild frontier, the true planet of war.

In the little rooms of the monastery Ada's sitting somewhere with the chief monk – Prior Stick. Who's giving her a cagey sort of welcome.

Ada, talking: *Prior, we know Bees from Mars, or one iteration of her. If you would just let us open a channel—*

The Prior, talking to Ada: *In good time, Ma'am Ada. Our patron is often slow to greet new arrivals. She is cautious. Perhaps if you explained to me why you are here. Cricket has told me some of it, but in a version somewhat fragmentary.*

And Ada just spilling it. Letting go all our secrets. And I don't know if they were any use as secrets, but I don't like just giving it away. Ada, saying Bees called us from Earth. Only maybe not this Bees, because we'd ended up next door at this Factory instead and wasn't *that* Bees? Ada isn't fitted with a camera, so I can't see that cowl-shadowed face, probably couldn't read it if I did. There's an interesting pause before he says, *Yes, of course. Bees has many faces.*

Suspicions deepen. In me. Not Ada, who'd swallow anything.

A snort, from beside Ada. The woman, Orme, who won't be peeled away from us, even though the monks don't want her there. The *old* woman, if she's to be believed.

She knows something, I think. She knows and the monk knows, and Orme could give it all away, and that's why they can't just force her to leave.

Ada, Wells, Tecumo, they were all old friends. I was on Crisis Crew, of course. Not bosom buddies with the mammals though. Wanted to come because... can't even recover that part of me. The Mars part. So let's just say it's the challenge, the hunt.

I'm hunting now. While you just sit there on the wrong side of a door and whine, Wells.

Ada is talking about the Griffins now, a subject on which she basically knows nothing. *We're so sorry. We didn't mean to cause trouble. What can we do?*

The Prior, on firmer ground: *I agree they'll come here. I will try to talk to them. I'm afraid the Bunkers are turbulent neighbours. Before, there were several of them claiming this region and we had fighting coming through most years, and multiple gangs of soldiers trying to tax Clearwater and the Apiary. After the defeat of the New Army, and the Dragons catching a sickness, well, the Griffins have maintained undisputed control of everything around here except the Factory. Which has the advantage that there's only one tithe, but the drawback that there's no curb on them when they want to flex their muscles.* And I guess Ada must have looked pretty damn miserable because he tells her, *They were going to turn up at our door and seek to alter our arrangements sooner or later.* As though it's his job to make her feel better. And he's wasting his breath because she's been misery since we landed and a few monkish words won't change that.

I am easing my way around the outside edge of the monastery building, following that whispery little trail I've picked up. At the back are the hives, I see. Actual hives of actual bees. No radio chatter, or not on any channel I can pick up. I creep closer. The bees swing past me and back. My camouflage is geared for human-standard eyes, mammal eyes. Bee senses register me.

Then it's you dragging at my attention, Wells. There is a monk come to talk to you. Daring to get within claw-range, jaw-range.

The monk, to you, Wells: *Bioform, is there anything I can get you?* Her voice shaking very slightly, breath uneven, but daring. I approve of daring.

You, to the monk: *I am Wells. My name. Wells.*

The monk: *I'm Siblen Boatman. Would you eat, or...?* Probably wondering what a huge monster dog would want to snack on. Monks, probably.

You, to the monk: *I am sorry, I am... having difficulties.*

You have a camera implant and I patch into it with one of my eyes. The monk – they have their hood down and their high collar folded back into a mantle over the shoulders, but if your nose didn't tell you *woman* you'd never know it from her face. She looks at you, and then – you're surprised, I'm surprised – sits with her back to the wall, down on the floor beside you.

The monk, to you: *Tell me. Perhaps we can help. Bees knows many things.*

I try to force my suspicions down the channel to you, but you are at a low ebb, Wells. You've been better. You're talking, and I'm only glad I'm not sharing your actual skull, just your eyes, because the rush of shame and guilt and weakness is probably a lethal dose for anyone who hasn't built up the tolerance you have. And like Ada, you just *tell* these people everything. All your sufferings, so that if they want a knife to use against you, you've put it in their hands.

I can see it's all going to come down to me, basically. You're rubbish, both of you.

And I'm looking for Bees. Not *bees*, which I've found, but big-B Bees. Is she here, within these hives? Are these little units, harvesting nectar and ferrying pollen, actually part of the great old Distributed Intelligence, or that part of it left on Earth. And if so, is the Factory also Bees or a different Bees or not even Bees?

I try to open a channel to them. *Are you there Bees? It's me, Irae.* But there's no connection there. Absolute dead air.

And something.

The whisper. The little voice that's trying to go beneath my notice. But running cold like this lets me focus. Not like you, Wells, with the storm of the world up your nose and in your ears. Run cold and I can focus very narrowly on things. While at the same time I wouldn't notice if someone trod on my tail until a minute had gone by.

Sometimes slower is better.

I give myself over entirely to the radiosphere. On Mars it's busy. Lots of people with open channels, and all the essential systems ready to report or sounding alarms, plus public information, group chat, and on and on. You could drown in information, on Mars.

Here on Earth: blank space, dead air. The faintest of scratchy echoes that might be distant long-range transmitters from hundreds of klicks away, or just random fluctuations of solar radiance and Earth's magnetism. Wells and Ada, strong channels, recognised and tagged and filed away. No Bees-talk. Not even the edges of it. No encrypted chatter of the units of a DisInt mind linking together.

That whisper.

And it's got an odd feel to it. I construct it in my mouth-imagination, feel it out with my tongue, that has more sensory cells per square centimetre than your nose, Wells. It's a channel, transmitting and receiving. I shift about, sending the bees looping past and around me, triangulating.

From outside the building, talking to someone inside.

Bees?

Not Bees.

A spy.

I dial up, not full goblin but I'm going to need some response time. I am instantly keen to *act*, in the mood to get stuff done when before I could just have lazed forever. Let's burn off some of this *heat*.

Wells' channel: *Irae, what—?* Aware of my sudden spike of activity, even as she's spilling all her secrets to Siblen Boatman.

Ada's channel: *Irae, no – no violence. What are you—?*

I'm done talking, though.

The spy registers me as I come in. Not with her eyes, but she's good with channel reception, catches some bleed of my link to you and Ada. A woman, just sitting like a beggar by the walls, and there are monks in sight of her. Probably they gave her the crust she's gnawing. They don't know. They don't know she's a live wire, carrying on an encrypted conversation with someone on the inside.

I go for her. She's on her feet, running, but I'm dialled up to nine which is like lightning. Earth's warmth and my vest's electric heat and *got her*. Jaws on her leg and then body whipping about her. Monks shout and scream and run. Bees quartering the air, but dumb, so dumb. Ada and Wells yammering at me, saying, *What did you do what did you do Irae what—*

I go up the wall with her. An excess of energy. Bracing. Need the exercise. Up the wall which is pitifully easy to scale. Up to where the wooden parts are, where the windows are bigger. In through a window. Shut window. Splinters, shutters, all pieces. Woman in my jaws shouting, screaming. Little fists beating at my scales, trying for my eyes. I shut all my eyelids against her and she can't get in. Not like me. I'm *in*, with her. Dragging her squalling down a corridor. Monks coming out to stare. Seeing the half-there-half-not lizard with its prey. I laugh at their faces but probably it doesn't come over as a laugh.

Drag her down some stairs, belly-scales against stone. Into the lower part of the monastery, hunting Ada. See you, Wells. See Ada. Burst in on the meeting. Wells follows behind me, whining.

My channel: *Caught a spy.*

Ada's channel: *What have you done?* Redundant, I've just told her.

I see the Prior there, staring, horrified. I see the woman Orme. Knowing, sly, rather delighted. A mean-spirited, nasty woman, I think. Maybe we will be friends.

And then, just as Ada's about to really explode on me, I tell her the other thing, the bigger thing. It's about time we had some truth around here.

37

THE WITCH

My self within the Apiary goes by the name of Siblen Darter. She works in the kitchens there. Under normal circumstances a surprisingly good place from which to know what's going on. Word always filters down to where the food is made and the warmth is. Not water-cooler chatter but stove chatter, especially with the chance of leftovers thrown in. All the monks go there eventually and Siblen Darter, herself, the individual, keeps open ears. And when I'm near enough and we establish a channel, she and I share a selfness. Thanks to the metal button we use as transmitter/receiver. Thanks to the fungus among us. And from time to time I leave her a packet of the mushrooms I gather, to renew her own inner network. Because it's not a species she'd want to be seen gathering, having no medicinal uses and, in fact, being quite toxic unless you're one of *me*.

When Siblen Darter and I connect and become one, I'm just in time, because the visitors have arrived. Arrived with Orme, and seeking Bees, and that's a powerful combination for trouble. My occasional co-gossip Jen Orme must be having the time of her long, long life. Or at least, of that stretched out part of it that post-dates the General Collapse. Short of entertainment for her, these degraded days. And while I have gone from being one sort of fugitive to another, she used to have every possible convenience and advantage at her fingertips. I think it was that which did for most of the Old Folk. That they could remember the wonders so clearly, and how much their word carried, how much control they had. Orme and Griffin and all the other immortals, they always thought that the end of the world wouldn't touch them. That the pond would shrink but the relative bigness of their personal fish would compensate. But when the pond dried up to a muddy morass they ended up flapping about gasping with the rest.

I, of course, consider myself an amphibian. I had already future-proofed myself – my multi-faceted self – against the end.

Most of those Old Folk are dead. Some by their own hand because a world where it simply wasn't possible to have all that stuff and influence and adulation wasn't one they could live in. Some settled for the rough and ready influence and adulation this latter-day world has to offer but even their sociopathic self-regard exhausted itself. Eventually someone takes up a sword or some old gun or just a heavy piece of wood, and that's it for you. In this world – the world that people of their stripe were so instrumental in bringing about – you can't hide behind lawyers or money, and none of them was quite able to build an empire or cult or warband with the requisite blind multigenerational faith. Stripped of the structures that enabled them, none of them were quite the genius demagogues they'd reckoned.

Jennifer Orme's own sociopathy comes and goes. She's aware of it. She puts it on a leash. One day it'll get the better of her and the people round here will beat her to death with a spade, most likely. But so far she'd threaded the needle of her own character defects, long enough to see the Martians return and royally fuck everything up for everyone.

In the kitchens, Darter doesn't hear the full details

of what's going on up in Stick's little office. A shame. I imagine the old Prior is sweating like a pig on meat day, however mannered his words are. There's a figure of speech from before the General Collapse I'd like to introduce him to: *House of cards.* A lot's about to come down around here.

I should probably move on before then. Don't want to be here for the fireworks. My network – my far-spread selves, depend on mobile nodes like me to keep making contact. Without that we're just... selves. Individuals. Which was never enough. Elsha in the Griffin Bunker and Darter in the Apiary and all my othernesses, we are too far spread to be *me* together, and so I have to take my *me*-ness on its endless pilgrimage between them.

Darter's ears, and her memories, complete my understanding of the playbook so far. Martians, for sure. Orme will remember them and so do I, although in my case these are the memories of other bodies passed down the chain of me that extends deep into the past.

So what is it I know? The Martians say they're here on Earth at Bees' invitation, hence they've come to the Apiary. Which is interesting, in that *house of cards* sense. The Martians have visited the Factory already, which makes sense. The Martians have already managed to get one of themselves killed. They've had a major run in with the Griffins in which they've managed to implicate the monks. The Griffins are mustering their full force and will be on their way here. Amazing. They've not been here two days and already they've destabilised this entire region. And while I don't necessarily hold a brief for the Apiary, it's a useful place. They have records that my self here can access. They've been a reasonably good neighbour, even if I've always been leery of them in case they do the relevant math and figure me out. And now maybe they're going to get wiped off the map because of meddling Martians.

I'm just setting some parameters that I – Darter – will follow after I – the witch – have moved on, when I register something nearby. An electronic hiss about the edges of my personal channel, like static. A glimmer in the air, like the sun catching on dust between me and the landscape.

It strikes faster than I can register. I still don't see it, even then. Jaws clamped on me, double-rows of pain and a monstrous strength of grip that speaks oh-so-eloquently of how little of its true force it's deploying. It yanks me clean off my feet. In the kitchen I – Darter – have fallen down, screaming. They think she burned herself, then they think she's having a fit, call for the apothecary. I am ready to be finished here – this old body, this wanderer. The monster has me. I'd be mad with panic if I was just one, but my mind is with Darter as much as here. I let the mind-curdling cocktail of fear and pain pool in our limbs and our gut while my thoughts keep clear between our heads for as long as my ability to transmit persists.

It hoists me effortlessly, the strength and speed of it is horrific. A physical heat comes from it, shivering the air, and I can still not make it out. A coiling, serpentine shape – felt by its pressure about me as much as seen in the eye. The colours of the world smear and bleed across it.

Oh god, I've seen this before. One of the old Bioforms, the reptile-forms, the dragon-models. Killers, assassins, black-ops monsters. And yes, that's me stereotyping a breed of *people*, but it's not exactly doing much to dispel the slurs right now.

We go in through a window in a hail of splinters. I hear monks screaming. I'm letting my limbs be loose, just go with the flow because I do not have the strength to fight against it and that way lies dislocation of the joints. Should I survive, I'll feel every bruise the moment I don't have any other head to flee to. Should I survive.

Then I'm on the floor and in company. The monster has brought me to its mistress. I see the other Bioform in the doorway, the one that's properly see-able. I see monks. I see Stick. I see the Martian woman, leaning against the wall as though it's *her* having the bad day. Everyone's shouting and mostly at the glass snake thing that brought me in like a cat with a mouse.

Except Orme, who's watching me like she just took over the mantle of cat. A little bit of a smile there, at the discomfort of her fellow tea-drinker. Orme is on the swing back towards sociopathy and this is bringing it to the fore. Not someone who's about to speak up to save my hide, I suspect.

"What do you mean a spy?" The voice of the Martian woman is shrill.

The reptile drops its mimetic camouflage. There's a fresh bout of shock and wailing from the monks. It's muzzle to muzzle with the other monster, which is probably some hyper-specialist dog-form but looks just as much of a nightmare. I think they're about to go for one another right there, and the sheer physical explosion of it will probably kill everyone else via collateral damage. A pair of real nasty-looking customers. They breed 'em ugly on Mars.

And then, from the Martian woman. "What do you mean there's no Bees?" A despairing cry loud enough to silence everyone, to part the two monsters from their incipient throw-down. And, into that quiet, just the mean old chuckle of Jennifer Orme.

38

ADA

It was the face of the chief monk, the Prior, Stick. Otherwise Ada would just have bulled on along her preferred version of events, the one where the monks were wise and honest and Irae was wrong. She saw the words land, though. Worse, she saw the reaction in the other monks, the younger ones. Bafflement, offence, worry. *They don't know*, but Prior Stick did. Knew that what she'd just said – echoing the accusation on Irae's channel – was true.

The Prior started barking out commands, sending them off, getting them to prepare medicine for this poor old woman Irae had just dragged in under the mad pretence she was a spy. Getting them to fetch food, to keep watch for the Griffins, anything to occupy their minds. All except Cricket. Cricket, who had Ada's elbow, not to hold her prisoner but to hold her up. Cricket, who was plainly as clueless as the rest but whom Stick obviously believed needed to be let in.

The old woman should surely have been sent to the infirmary or whatever they had here. Irae hadn't actually torn her up – the Bioform had hinged teeth and a lot of control over just if and how she jabbed them in. There was some blood, though, and surely a great deal more mental trauma. That outward calm must just be the symptom of a paralysing shock. Except the woman, Orme, stepped in and said the newcomer should come with them, and Stick agreed. Something was going on that Ada was excluded from. This old, hurt beggar-woman was known to both of them. Was something other than just a random victim of Irae's spiralling aggression.

So it was that the Prior's little office became a very crowded place. He and Cricket, as representatives of the home team. All three of the living members of Crisis Crew currently resident on Earth. Jennifer Orme, and the new woman, as visiting locals. The *witch*, as she was referred to by both Orme and Stick. Not an insult, or not just an insult. A... category, perhaps.

"A spy," Irae said. "An open channel. And someone on the inside."

"Nobody here has channels," Ada insisted. "Not outside this Factory place." A moment's doubt. "Unless Bees..."

"There's no Bees," Irae insisted, aloud, obviously relishing the Prior's flinch and Cricket's increasingly strained blankness.

"Interesting," Orme said. "I mean *I've* technically got the implant for a channel. It stopped transmitting some time ago but I still receive ghosts and whispers from it now and then. Just like I'm doing from you right now. It was state of the art when I had it put in back in the Old, but some damn thing's gone wrong with it. Just have to hope there isn't a tumour building around it or something." She gave a raspy little chuckle. "But she has one, does she?"

"Yes," Irae said, and "No," Ada said, but then felt just the edge of a connection even so. The faintest suggestion of electronic *presence*.

"Who is she?" Wells asked. The pertinent question.

"Not entirely sure," Orme said. "The latest in a long line, though. These witches have been around since forever. A sisterhood. You want to talk about preserving knowledge of the Old, then don't bother with Stick here. He has a bunch of records, sure, but she *knows*."

"She's like you?" Ada clarified.

"She is not," Orme said firmly, but then shrugged. "Probably that's for the best. I don't think the world needs more people like me." Sounding abruptly disgusted at what that was. "There's always a witch come calling. Not the same face, over the years, but the conversations just pick up where they left off. I have some theories."

A monk came in then with some ointment, cloths, water. An awkward intermission when they cleaned the punctures made by Irae's teeth. The witch – not so old, Ada saw, but dressed to seem older – watched everyone very narrowly, wincing at the attention.

When the medical monk had retreated, Wells said, "Bees."

Prior Stick took a deep breath. Looked at the witch. "Perhaps—"

The woman scowled at him, then at everyone else, a different shade of dislike for each one of them. "If you're about to tell them your grand insect-related secret," she said tartly, "I *know*. Even if my channel wasn't telling me of the glaring *absence* around this place, I've always known."

Stick sighed. "Siblen Cricket, I need to rely on you. To be the Apiary's ambassador to these visitors. To go with them, perhaps. To assist them."

"I understand, Prior."

"In which case you should know the truth, seeing that we can't hide it from them."

The look of worry and pain on Cricket's face right then made Ada wish she hadn't said anything. But it was said, and the Prior had chosen to bring her protégé in on the secret, and so…

"No Bees," Ada said again.

"Not a whisper," Irae confirmed. And it had a better radio suite than she did.

Orme was smiling blithely at the Prior's discomfort. But then she'd been here from the start, perhaps seen the place founded. No secrets from her.

"In this place we seek to preserve everything we can of the Old," Stick said. "It is our mission. For when such learning might be useful. Or simply because to preserve is better than to destroy. The world out there is built of sand. Fleeting lives and forgetfulness. But they remember Bees. One of the great powers of the Old, the mind of a thousand fragments. And Bees was lost with the rest of the Old, but we preserved their name and reputation. The memory of Bees survived into the New, and we hid behind that reputation. A secret passed down from Prior to Prior to me. Our protector, who would strike back against those who harmed us."

"That can't possibly have worked," Ada objected. "You're saying nobody ever…"

"Oh they did. They have. Our people, robbed, murdered, cheated. But we invoke the name of Bees. They remember, the next time events turn against them. The hand of Bees working invisibly in the world. You'll hear many stories of those who crossed us and met some dire fate. Bees can be everywhere, do anything. It is… a convenient belief. I would like it to remain so." A look from face to face, acknowledging that they could seek to ruin the Apiary, strip it of its mystery, should they choose.

Cricket was looking absolutely stricken. Ada wondered how many times he had blithely relied on that illusory protection.

"But you are from the Mars colony," Stick said. "Or so you say." Something he must have read about, piecemeal, from ancient records. Stick was learned, and Orme just plain remembered back when the idea of a Mars colony was commonplace. The witch... Ada saw that eerie calmness in the woman, almost a disconnection. And displaying no surprise whatsoever. Ada's mind was beginning to skip, robbed of certainty every time she tried to trust her weight to the situation. The witch, Orme, Bees... too many parameters of the world all shifting at once, tugging her in different directions. She should be taking command, doing something decisive like Tecumo would have done. Instead she just wanted to curl up and weep. It was all too much.

And then the Prior said, "But you have met Bees?" And that was a whole extra complication because there was a genuinely religious reverence in his voice. As though the lie the monks had told over the generations had alchemised into a real faith. That Bees was out there, and might come to save them for real.

And Wells burst out, "Yes, and we've brought—"

Ada's channel: *No, Wells.*

Wells' channel: *But we brought – we should—*

Ada's channel: *We might. But it's...* Complicated. A terrible idea that she had already made up her mind to pursue. But she didn't want to get into a debate with the Prior that was plainly going to enter theological territory. There would be time, later, to explain that, if he really wanted, they could make his Big Lie real.

"Bees is on Mars," she said firmly. "She doesn't interfere with us. But she's there. After the signal, we contacted her."

She saw blank looks, though, and of course she'd just introduced a whole extra element that none of the locals was familiar with.

"We received a signal," she explained. "That's why we

came. That's why we travelled all this way. A signal from the Bees on Earth, asking for our help."

Watching the Prior's face, seeing it twist with a dreadful hope, but also fear. *What if Bees is real, but isn't on the side of the monastery? The worst of both worlds, surely.*

She wasn't going to say any more, but Wells just blundered ahead of her, saying, "The Bees in the Factory. What you call the Dog Factory."

They had not known. It was apparently the big day for everyone's secrets to be dragged out into the air, and Ada wondered whether the delicate balance of this part of the world would ever recover from the Martians' well-meaning interference. The Prior, wide-eyed. Orme startled, discovering that even she hadn't known everything. But the witch unsurprised, nodding. And if she had a working channel, and if she'd gone up close to where the Factory was, then she'd have picked up the chatter between the people there. The dogs and the humans all connected just like in the old days. Suggesting a perspective like Orme's informed by a real knowledge of the pre-Crash world...

"Who are you?" Ada asked her. "Really."

The witch met her gaze flatly. "Funny way of saying you're sorry your monster mauled me."

"Not her monster," Irae snapped, even as Ada opened her mouth for the apology. "My own monster."

The witch's lips twisted into a thousand-year-old smile. "Well we have that much in common then. I'm my own monster, too."

No clues from anyone else there. Stick and Cricket blank, Orme hooded. If she knew the truth of these witches, she wasn't sharing.

Some order, like the monks, preserving knowledge down the generations. The obvious answer, that Ada knew wasn't true. But the witch met her imploring gaze and just shook her head.

"Let Orme blurt out her provenance to every new visitor who comes to Clearwater," she said. "Some of us keep our secrets. Now either have your monster try to tear it out of me, or let me alone to heal. You've done enough already."

Ada thrust her hands into the air – an insuperable effort, it seemed like – and almost shouted, "We didn't come here to make trouble. We came here to *help*!"

Into the quiet which followed that, Orme chuckled. "Tell that to the Griffins," she said.

And the quiet, after that single bitter snapped-twig of a line. The quiet that was the echo of too many revelations. A tower of broken secrets piled up over generations and knocked down in moments. Like the quiet after an explosion, when it's just the dust filtering down. Wells shifted – the biggest thing in the room, her shoulders close to the ceiling, her back to one wall, so that everyone else seemed to exist in the concavity of her arched body.

Irae snickered.

Ada's channel: *Don't you—*

Irae's channel: *Oh grow up. What does it matter?*

Ada's channel: *Everything we do makes things worse.* She had a moment of fiercely hating dead Tecumo then. Because he'd always been so positive, so sure everything was for a good cause. And see where his enthusiasm had got them.

She looked at Cricket, who had the sort of expression someone got when their family was killed in front of them. And the actual killing was still to come, but Ada had also turned up and killed his idea of how the world worked. Showed him how this parent figure of his had been lying to him. In his return glance was everything anyone would need to know about him. Gaunt with poor nutrition, young, desperate for something to believe in. She wanted to ask him what Bees had meant to him, the belief that there was a Bees within the Apiary. She remembered him using the name to conjure with. Bees was his totem, the thing that he *had*, to balance against all the world's ills.

"Prior Stick," she said, eventually tugging at the man's sleeve to get his attention because her voice was weak. "I need to talk to you alone. Really, just you."

Wells' channel: *I'll be ready.*

She couldn't even bring herself to reply, waiting until she had the room just to herself and the Apiary's leader. Orme was last out of the door, protesting, but now the truth was out she had no blackmail material left, and Wells hulked at her, and she left.

We are about to do another stupid thing. And Ada actually spent a moment trying to work out if that was Wells or Irae in her head, then understood the thought was native to her.

"You do not have Bees here," she said.

He shrugged. "As you have discovered."

"But you've been telling people forever that you have Bees."

"I don't know how it even started," he said wearily. "Perhaps the earliest Priors thought Bees would come here if we prepared a home for her. Or – we've always kept the regular kind of bees. Perhaps that grew into a rumour or a pretence that we had given a home to the intelligence. But that belief has protected the monastery, down the years. Allowed us to do some good."

Ada drew up her courage and her strength, both feeling in perilously short supply. "I can give you Bees."

Stick blinked, frowned at her. "Hmm?"

"We have, from Mars, a starter colony. A gift from Bees,

THE DANCE

to be used *in extremis*. If we woke it here, it would connect with your hives outside. It would grow and become Bees very quickly. And then you would have Bees. Cricket said you have a radio?"

"We have tried to maintain it." Stick's voice shook, just a little.

"Bees could speak to you. For real. And act in the world. It would be something."

The Prior closed his eyes. "There are many tales of the threat Bees posed in the Old. That she would have enslaved everyone, taken over the world."

"There's a saying of evil men: always accuse others of doing what you are doing. On Mars we know Bees as an enemy of human powers who were trying to do just that."

The Prior took a deep breath. "Will she shield us from the Bunkermen?"

"She will need time to grow and come to an awareness of herself. So not right now, but later. If we can hold out this time."

Stick sighed, a man with no good options. "Then yes. Let us become what we are supposed to be. Or at least have a chance to."

Ada's channel: *Wells, do it.*

Later, Ada finally had some time to think, with Wells still nursemaiding the process at the hives and Irae – somewhere. Honestly right then Ada didn't want to even *know* where the reptile-form was skulking. And Cricket was being brave, she supposed. Obviously deeply unhappy, desperate to find a way to bandage the wound that had just opened up in his world. Not spilling the beans to the other monks. Not telling them they'd been deceived. Just as she hadn't told him that the lie he'd been let in on might be about to become the truth. Even though she was doing it for him, in the end. Not because it was right or wrong, and not to further some Martian agenda, but because Cricket was a good kid who needed something greater in the world, and she could give it to him. If any of them survived. But, because the knowledge would only screw with him further, she didn't tell him. Cricket remained thoroughly miserable, therefore, and made a poor job of putting a brave face on it.

Ada wasn't happy herself. Sitting in the monks' infirmary, leaning back against the wall because she could barely hold herself up. Helpless. A liability. Worse than that, because she'd just blundered into this world, they all had, and made stupid mistakes that had already killed Tecumo and were just getting going with the repercussions.

In Tecumo's dream of Earth they'd arrived open-handed with gifts of goodwill and knowledge, and the grateful people of the homeworld had laid out their histories and welcomed the travellers from afar. Honestly, with that amount of optimism it was amazing he'd lived long enough to leave Mars.

She linked to Wells, out with the bees. The starter hive was installed there, the first generation of Mars-grown insects already hatching out and making pheromonic contact with the native Earth swarms, integrating themselves, hacking the local system. Wells sat in the midst of the busy-ness of the place, the bees looping past her in patterns that shifted subtly as they became more than they had been. *I do not know if this was the right thing to do.* But Ada had grown up trusting in the benevolence of Bees. Like the monks, she felt as though she'd been left with nothing else to have faith in.

THE DANCE

One of the monks was out with Wells now. Offering something. Honey, Ada saw through Wells' camera. She almost laughed at that. Actual honey, as locally sourced as you could possibly get.

The monk – Siblen Boatman – was explaining something. Allergens, herbs, some manner of preventative remedy. One of the Apiary's exports to the other communities around. Based on 'writing from the Old' and no way of knowing if it was actually true or just another strand of the fragile protective cocoon the place had woven about itself. It would help Wells, Boatman claimed. A side-effect, really. Dulling the senses. A caution about possible side-effects.

Tentatively Ada put out feelers to see if Irae was nearby. After several failed attempts to link, she found the reptile-form outside, watching the horizon for the Griffins. Marking out ranges on the landscape in her head, probably. And it was Ada's duty to intrude and tell her not to kill anyone or make more trouble. All that responsible stuff. Ada, the sensible one. The dull one. The one who tried to help but, let's face it, wasn't actually of any use to anyone.

She said nothing.

Wells and Irae would be able to defend the Apiary to an extent. Irae, gleefully, Wells with reservations. And it was patently supposed to be Ada's role to simply *be defended*. And she'd had enough of it. Enough moping. Enough giving in to the world. Enough weak words.

At her request, the monks brought Cricket to her. He still looked aghast at… everything really. The bad things that had happened and the bad things yet to come. Like her, he was falling, scrabbling desperately for purchase on a world that didn't care. She was about to offer him that purchase. Because she needed to act but she couldn't do it alone. And Wells and Irae would probably try to stop her.

"The Factory," she said. "Your Dog Factory. They have Bioforms there, a good number. Tecumo reckoned forty, fifty?"

"Maybe that many, maybe less," Cricket told her. "They don't let you very far in."

"Even twenty Bioforms would be enough, to push back the Griffins. They see a line of dogs waiting for them, they'll probably turn round and go home."

Cricket said nothing.

"You said the Factory fought off these Bunkers before."

Cricket said nothing.

"They owe us," Ada told him fiercely. "We helped them, and then Tecumo got killed after. They owe us help, for what Tecumo did to sort them out. They owe us and you know them. You monks, you're part of a system they use. That's what you said." Or it was how she'd interpreted the scraps of information he'd blurted out to her. "We're going to the Factory, just you and me. Because I think neither of us are any use just sitting here behind walls. We're fetching help. We're saving your people. Are you with me?"

Cricket said nothing, and his expression did not measurably improve, but he nodded.

39

CRICKET

The Martian woman wanted to leave swiftly because her friends would try to stop her, Cricket divined. He wanted the same thing because Prior Stick would *certainly* want to stop him. Because what was Cricket, exactly, to accomplish anything? An anonymous messenger, the agent of others' decisions, no more. Looking at Ada, slumped in the seat of the Griffin car, he had an odd sense of kinship with the goblin-looking woman. She was weak, sick with just being on this world. The other two, the monsters, were strong and murderous, like elemental forces. Cricket felt he understood. Ada and he, both of them, had made this mess. Had put the Apiary under the shadow of the Griffins. And then been left out of any plan to fix things, discounted as liabilities.

The Apiary monks walked. A generation ago a couple of the villages had possessed ancient sun-powered vehicles, he'd heard, but they were just rusting landmarks now. When Josh Griffin III had built his bunker at the end of the Old, though, he had done his best to future-proof it. The place still had a fleet of cars that ran, even though they were decidedly more patched and jagged than they had originally been. With this stolen vehicle, the journey from Apiary to Factory was going to pass by absurdly swiftly, barely a few hours' jolt over the rough track.

"We did some work, on the car," Ada said. "When Irae

was off... fetching you. Most of its systems had been offline, all the driver assistance stuff. And we could do that, for them. We could help fix things. I mean they live in a bunker, dug in. Half the problems that causes, we have on Mars too. I could tell them I'll sort their ventilation and temperature regulation. If we can only talk to them. That's all I want: a détente. Enough of a threat behind me that they'll stop long enough to talk." Her voice coming in fits and starts, jumping with the car's suspension, wheezing as she dragged in long breaths.

"I don't understand how the Factory can be Bees," he said, a little later.

She spared him a glance, then had to wrestle with the wheel to keep them on course. "Bees isn't God, Cricket."

"What?" he blinked at her. "I... know that."

"Do you even do God, at the monastery? In our records a monastery is... a God thing."

"It's not encouraged," Cricket said, clutching at the roll cage to keep his seat. "Bees..." A stab of pain. "Well the Prior said Bees wouldn't like it. That Bees was the science of the Old, that we could believe in. And gods and superstitions weren't helpful. Obscured the truth." And a ghastly, wretched laugh squeaked out of him, because it turned out that the thing he'd believed in most of his life had, after all, just been an object of misplaced faith.

"I'm sorry," Ada said hollowly. She took a moment to gather some conversational strength and then told him. "Back in the early days on Mars there was... like a mystery cult, of Bees. People secretly doing what Bees said, back when people thought Bees was the enemy. And there are still some sects, you know? There's a weird thing some of the dog-models have, and other people just have God, of one sort or another. But it's a bit... embarrassing. To people of my generation. It's something you can do, fine, but if it gets in the way of the work being done, or if it means you start telling other people they have to do or not do what your particular flavour of teaching says, then you get stared down hard by everyone else. I mean, honestly, your monastery, Hell City, I think there's overlap, in the way they work. Disaster communities, no room for it."

Cricket considered that, and how some of the Bunkers had gone, elevating their way of life to a divine mandate and themselves to ordained lords of creation. The Griffins were far from the worst of them, perhaps even the best option to have won out, given their relatively mundane demands on their neighbours. He wasn't sure that being placed in a disaster situation necessarily made people focus on the worldly. And Bees had been a lie, after all. A genuine 'religious figure tells you a huge thing about the world that turns out to be utterly false' level of lie. That fell in with God stuff, didn't it, even if successive Priors had dressed it all up as science?

"I never saw any bees at the Factory," he complained. "Bees, the insects. Or Bees the... the Bees. There are the people and the dog-Bioforms."

"Bees is information," Ada said. "She was actual bees once, and she's returned to that time after time. Probably there'll be a hive at the heart of it, but she might just be virtual, trapped within her network. That was part of what we were helping with. The Factory has some very sophisticated data architecture. Like this Josh Griffin, Bees was thinking ahead. Except most of it was glitched, disconnected or misconnecting. Degraded over time, abandoned as Bees contracted. A lot of the complex functions it was designed for were dysfunctional, broken links in the logic. And the Bees we know on Mars could have sorted herself out, for sure, because she's got space to manoeuvre. But here I guess it would be like trying to do surgery on your own brain in a cupboard. So we came to help."

By then, they'd been passing through the Factory's farmland for a while. Not quite the course that Cricket had trodden from Brokebridge with the cart, but converging on it. The walls of the Factory were ahead, the fortress that had thrown back the forces of three Bunkers, a generation ago. The moment that had set this part of the political landscape in stone, and left the militias to bicker and jostle amongst themselves.

They were close to where he'd found the body, of course. Not something he felt she'd thank him for mentioning. Close where he'd found the body, and where the dog-Bioform had found him. Where the ruin of his world had all started.

"They don't like random visitors much," he cautioned Ada. "But I should be able to talk to them. I know them, they know me. They know the Apiary. We're the go-betweens, because the villagers don't like coming here. Scared of the dogs. They'll let us in. If I ask them." Feeling the burgeoning rise of hope in him that, yes, here was something he could do to *help*. He, Cricket, least of the Apiary monks.

"They'll let us in," Ada agreed, but then he realised she wasn't just agreeing. It was an independent statement. "Bees has opened a channel. Recognised me." And, indeed, as the car approached, the gates of the Factory opened. Stern faces, human and dog, watched them approach.

"But I thought you... I thought I was guiding you. That you needed me to..." Cricket stammered.

"I need you," Ada confirmed, and then dashed all his dreams of significance with, "to help me out of the car."

★

He saw a welter of different emotions on the faces of the people who met them. Bafflement, curiosity, alarm, and something in Bellman's face that Cricket couldn't read at all. And the dogs, pressing close, whining faintly, eyes seeming faintly agonised, as they always did. Just the Factory, and if the Factory was the most uncanny place Cricket knew, that was because it was of the Old. Like Josh Griffin and Jennifer Orme, like the knowledge the Apiary had painstakingly collected. An unthinkable time when human beings had reshaped animals and put people on other worlds.

"Siblen," Bellman said. His eyes were wide, brows furrowed. "These types." Focus skewing over to spear Ada, who was doing her best but still leaning heavily on him.

"Friends, Bellman. Come to help Bees. But now *we* need help," Cricket said. But of course the man already knew the Martians were friends. Their other one, the dead one, he'd been here, giving just that help. The initiating event that had brought everything else, like opening a tiny door on a great flood. Bellman's taut expression didn't seem to admit that friendship, though.

"Talk to me," said Ada, to neither of them. Bellman stiffened, jolting straight as though bitten.

"Come," he said, sounding slightly strangled, "to the terminal room. The signal is strongest there."

A faint incongruity, as they were let through the doors of the Factory proper. *I thought their signal could reach all the way to Mars.* So what did the Factory need so strong a signal for now, just to talk to Ada. Cricket frowned at the thought. But what did he understand about it?

40

ADA

She'd not spoken to Bees direct, on Mars. Just seen the transcripts, had the conversations recreated for her second hand. She'd been expecting a conversation over comms direct to the channel she could detect, identifying as *Factory Admin*. She could distantly detect it talking on encrypted channels with the humans and Bioforms around her, but instead of keeping a channel open to her, its human agent, Bellman, was relaying its words. Cagey, or perhaps it was still repairing the accretions of time that Tecumo had helped it circumvent. Nonetheless, here, she knew, was Bees, the legendary, the original. Earth-Bees, from whom the saviour Bees of Mars had sprung.

And yet couldn't save Earth from the Crash, she thought. But by the time the Crash happened, Bees had washed her myriad tiny hands of Earth. Bees had relocated to Mars, save that this local branch had never gone anywhere except burrow down here in this forbidding-looking place.

The room they took her to must have been where Tecumo stood. These people here were some of the last to see him alive. And instantly she wanted to make it all about that. She wanted to stand in his shoes and see what he did, as a surrogate to actually getting to say goodbye. But that wasn't why she was here.

Again that channel, *Factory Admin*. She opened her

own comms to it, waiting for the electronic voice of Bees to buzz words into the small bones of her ear so she could hear them in her mind. There was a data handshake, then silence. Instead, the human attendant, Bellman, twitched as though prodded with a stick, and said, "You have come about your friend." Second-hand communion with the invisible godhead.

Yes. "No." Cricket's eyes were on her, mute but pleading, like a dog's. "We helped you. My friend. Tecumo."

Another twitch. "Yes," said Bellman. He was staring at Cricket – at both of them but mostly at Cricket. Eyes fierce and urgent, mouth moving slightly when Bees' message wasn't co-opting it. As though trying to open some channel between them by will alone.

"He's dead, Tecumo, my friend." Ada sent the words over the channel, too. Better to talk that way, surely. And *not* only because, that way, Cricket wouldn't be able to eavesdrop. She'd needed him to physically get her here but now his presence felt like being in a conversational minefield. She and Bees needed to be able to talk freely about things that might shock him.

"Yes," forced out of Bellman. The big dog that had stayed with them, the one called Deacon, whined very slightly.

"The people who killed him are going to come and attack the monks," she said. Still that open channel, still data flowing, but apparently it had to be this way. Just wind and words. "They attacked us." She'd been about to say *It's our fault*, but was it? They killed Tecumo. They attacked first. Wells had just been protecting her. A surge of righteousness came to her. *They deserved all they got.* "They kidnapped Cricket, this monk here. So we went to get them back." The thing she'd been furious at Irae for. That had made everything so much worse. But now it felt *right, just*. They'd only been reacting to the tyrannies of this jumped-up pack of thugs, after all.

There was a sudden click, around her jaw. A physical, painful jolt as though a sinew had snapped. She hissed in pain, waiting for the implanted endoskeleton to betray her yet further. Instead, what flooded her was a weird certainty, a crusading fire. In an instant she was no longer weak and she could feel her exoskeleton systems writing and rewriting themselves, thrashing like a snake within her neural implants as they fought her body for them. She became strong, rather than this sagging sack of defective meat. The thought was clear in her, an utter condemnation of her weakness, a promise that she need suffer it no more. The siren song of *You are in pain. You are not functioning properly. Let me fix you.*

Please yes, she said on her channel, which somehow transmuted to *access granted*.

It was incredible, how certain she could be, after that. "These Griffins," she said. "They want a fight. We're going to give them one." The ridiculous words, but they felt very *right* in her mouth. Adrenaline – the fierce and fiery Earth type, perhaps, rather than the cold blood of Mars – surged through her. And hadn't she wanted to *accomplish* something, coming here. Hadn't she been desperate for significance. Not the strong one, not the deadly one, just useless, broken Ada.

Now she discovered that she carried a great treasure in her head – discovered as if the top of it had been lifted off so she could see inside. Her implants, the artificial complexities added to and growing with a Humaniform brain. She was magnificent. She was an alchemical marriage of the human and the virtual. She was a bridge between the singular mind and the distributed one. She was perfect.

"Ada," said Cricket, sounding concerned, which was weird given how good she felt. She realised her nose was bleeding. Red Martian blood on her lips, down her chin. "It's just Earth," she told the monk. "It's just me. It's not important. We're going to bring Bees to the Apiary, Cricket." The monks had always claimed they were the servants of the Distributed Intelligence, and now it would become nothing but the truth. And that would only be the start of it.

The human agent and the Bioform agent were talking with Cricket now. The focus of the Factory was entirely on her, though. She could feel flashes of connection all the way through her mind. Improvements, bridges, an architecture of new logic and connection being built into her. The brief wail of security systems fatally compromised, of tolerances exceeded. Modifications far beyond the intentions of those who'd designed her implants. You'd need to be a comms wizard to set up a system that could make scalpel-keen changes like this to an active living system. You'd need to be Tecumo, who'd come here and gifted the Factory with his expertise. Who'd opened the box and discovered…

In her mind, very briefly, was the thought *All the evils of the Earth*, but it was washed away, painted over as *True enlightenment, singularity, perfection.* Sweetness and Light.

An insect landed on her arm. She twitched to brush it off and then understood that it was hers, on her side. The ancient, dusty terminal was crawling with them, a handful of black and orange, armoured forms and more forcing themselves out through old, eroded seams. The first new generation of Bees, just as would be hatching out at…

She stared at the creature on her arm as it flexed new, drying wings. Not a bee, really. Not any natural species. Something of the hornet or the sawfly, or the ichneumon wasp. An arched abdomen like a question mark. At its terminus, not the thorn of a stinger but a long, stiff wire, like an ovipositor, but of what eggs…?

The Bunker would be arriving at the Apiary gates even now. The Factory would march to answer them. To reclaim the Apiary and make it the thing it was always supposed to be. And then…

A wrong note. A memory glitch. Hadn't she already…?

Ada's channel: *Bees is already at the Apiary.*

Factory Admin's channel: (Speaking to her as naturally as her own thoughts, another self in her skull.) *What have you done, Ada?*

Ada's channel: *We have introduced Bees to the Apiary already.* And her head had been so full of the plan of doing just that, that she was trying to work out if that was *actually* what she'd done, or if it was just something she'd thought of doing, or dreamt it, or…

But no. She'd told Wells to go ahead, and even now there would be a Bees mind cohering within the hives of the Apiary, instantly provided with ten thousand bodies to build her distributed intelligence around.

A sudden spike of alarm. Her adrenaline system shooting lightning through her brain. *Fight! Flight!* A knowledge she'd done the wrong thing. An incongruous thought of *Bad Dog!* so that she cringed back, inside. And then purpose filled her again, ironing away all dissonance of thought.

"The Apiary has become a threat," she declared. Not her words, not her thoughts, even though they formed inside her head. "We travel there now." Not to defend it from the Griffins, though the Bunkermen would rue the day if they were there. To reclaim it from the pretender. Now the Factory had been rescued from the chains of its corrupted data architecture by the Martians – by her Tecumo – there was only one power that could threaten it. Itself. Its other self. Bees.

41

DEACON

I stand there with the monk and the pale woman and Bellman and I feel myself falling. Not my body. My feet are on the ground of the terminal room. I have nowhere to go. My head falls. My channel drops and reconnects, drops and reconnects. Here in the heart of the Factory. Factory Admin is doing something complicated and I keep being thrown out of contact.

I whine. I try to signal Bellman, but I can't get through to him. He has been speaking for Admin to these visitors, but now he is cut off as well. His hand is on my arm.

"You're all right, Deacon," he says. His voice, comforting me. The first voice I remember hearing, when I was a pup. Before I went through all the pain and the cutting and became *me*.

"Bellman." My voice, the rough, wheezy one, not the easy connection of the channels. The air around us is too full of the Factory's voice for us to talk that way. That voice, speaking only to the small pale woman. Not to her; into her.

"Boss is just doing something, that's all." And I know that.

Every time my channel tries to connect, I catch moments of what's going on. The openness of the woman visitor, like her brain is out and being worked on. She's gone very still.

Her mouth moves. Things under her skin like rods and ribs move too, shifting how they sit there.

"Boss is fixing her," Bellman says to me. This strange communion we have, that is just voice and ear, and not the great space of the channel. I think if he was not speaking to me I would go mad.

"Fixing," I say, trying out the word. "Like the other one?"

"Not like that one," Bellman says. "I hope." And then I realise the other is still there, the monk. Listening to us speak with our clumsy voices. Staring at me.

"You're the dog who met me, when I was delivering," he says.

I whine at him. I should not be speaking to him. Every time my channel drops, I have the old relic command that all outsiders are enemies. Except he's right here and Bellman is not worried about him. Worried, yes, but not about this little monk. Worried about me and Admin and the pale woman.

"Aren't you?" the monk presses. He's right there, right in my shadow. I could crush him. My relic commands say I should, but Bellman... Bellman is biting at his lip, looking at the pale woman. He speaks in the gaps when our channels are down.

"Tell him," he says, and so I tell the monk, "Yes."

"What were you doing out there?" The monk is very scared. Of me, and of bigger things than me. "Because... there was a body. A man like her. A dead man. Right near the path where my wagon broke. Where you found me."

I whine, deep in my throat.

"A man who was here," the monk whispers, although there is nothing which cannot overhear him. Except Admin is busy with remaking the pale woman so there is only Bellman and me to hear. And Bellman is very tense and very scared and his hand is on my arm.

THE DANCE

"Tell him," Bellman says.

I have the moments in my head. I was there. Taking the little man. Killing him, one shot. Taking his body away. Spraying it and me with the stuff Admin made, which killed my scent and made me alien to myself. Not knowing the *why* of any of it, only that it was for Admin, and so it must be right. He was a threat. The little pale man who came to help. Somehow a threat.

I killed him. He came to help and I killed him. It was the right thing to do because Admin said it was. When my channel reconnects I know this and am at peace. When it drops, because Admin is still focused on the pale woman, I know it was a bad thing and I am a Bad Dog for having done it.

"Why?" the monk asks. I don't even need to confess to him. He sees it in me.

I just whine. I cannot make myself face him, even though he is just a tiny scrap of man, and I could tear him apart.

Bellman takes his arm. Even Bellman could tear this monk apart. But that is not what he is doing. Bellman is pulling the monk away. Out of the terminal room. Away from his pale woman friend, and the monk is too weak to stop him.

"Go," Bellman hisses in his ear. In the room behind us we hear the woman say, "The Apiary has become a threat," and the monk's eyes go wide.

42

WELLS

I'd wanted to go after her, when we'd found she'd gone. And she shouldn't have been able to *go*, but I was... distracted. I'm still adjusting to...

The monks have given me something. Honey and herbs and something. For allergies, they say. But it deadens the nose, the senses. Makes me numb. I'd told them my problems and they had... a solution.

I feel like I'm outside in the furthest reaches of Mars, where you have to wear a helmet still. Where the atmosphere's still too thin or at least too poor in oxygen. Less and less of that each year, back home, but there are still regions, isolated by weather systems, distant from Hellas Planitia. You suit up, helmet on, and the world becomes just a little ball around you. All your senses crushed in around you. And normally that's bad but right now it means I can *think* because all that hell of distraction is taken away. Think all kinds of things. It's like the hard gravity of Earth has been taken away from my mind.

Honestly I find myself thinking some weird things that wouldn't normally find a place in my head. Because parts of me are reacting quite wildly to some of the ingredients. It's a bit of a trip.

It means I missed Ada going, and Irae either didn't see or didn't care. And when I do finally work out what's up, there's a note waiting for me. A little recording she left for me to find.

I'm going to find help at the Factory. They have Bioforms there. You have to stay and hold them off. Talk if you can. Help the Prior. Anything. Just buy time.

My channel: *I'm going after her.* After I've brought Irae up to speed.

Irae's channel: *Oh sure. Go running off, Good Dog. Or is it Bad Dog, lets all the monks get killed?*

My channel: *Don't – you're not helping.* I am so angry at her, that *now* she has to be like this. Like Irae. And I want to say it's not Good Dog, Bad Dog, and that what I'm judging her on isn't good, bad, either. I am two centuries and change from being just *dog*. But the *dog* is in me and the words cut, and for a moment I want to find Irae and fight it to show it I'm in control, even though I'd be showing it that the *dog* was in control and not me.

My channel: *You don't tell me what to do.*

Irae's channel: *Nobody tells anybody what to do. Even Tecumo was just the one we listened to, not the one in command. Go if you want. I'll stay here and shoot Griffins.*

My channel: *Ada said not to fight.*

Irae's channel: *You better stay then because otherwise what am I for?*

My channel: *How many Griffins did you kill at their home, Irae?*

Irae's channel: *Ten, a hundred, a thousand.* It was running too hot to know.

I get as far as heading down the hill. The monks don't stop me – couldn't stop me, but don't try. They don't realise I'm leaving to go get Ada. Abandoning them.

I stop. At the very back of my throat is that whine, so high a human couldn't really hear it. The unhappy dog in me. Everything I could do is the wrong thing. Ada and Irae and Tecumo have put me here, where I'm letting people down whatever I do.

Irae's channel: (Snickering.)

It's only then that I work out Ada didn't go alone. I go to where the car was. My dulled nose means I have to practically press it to the ground before I smell the fading trace of her, and of the monk, Cricket, also. Is that even better? It means at least she has someone to lean on. Even someone as small and weak as Cricket.

I make to go after them two, three times. I'm not as fast as a car but I can run. I feel helpless, and understand that's how Ada must feel, with her body betraying her. So she has found a way she thinks she can help. And here, would she be able to? Or just be a thing for the Griffins to threaten?

There's a game, a mind game. Some human with sheep and wolves, crossing a river. Can't leave either side with more wolves than sheep but still need to get across the river. I never thought about it before. Because why does the human care about the wolves? Leave them on the wrong side of the river, surely. But that's not how it is. The wolves are important somehow. Maybe the whole child's game is a lesson on balanced ecosystems.

Irae's channel: *What are you talking about?*

In the stuffed-up quiet of my head, my nose not talking to me, I have been speaking on-channel to fill the quiet. Didn't realise I was doing it.

My channel: *The wolves are important.*

Irae's channel: *Seriously what the fuck, Wells?*

My channel: *In the game you have to save the wolves as well as the sheep. But you have to save the sheep from the wolves.*

Irae's channel: (After a pause.) *I mean as tactics manuals go it's not Sun Tzu.*

I am having some serious enlightenment times right now. The stuff the monks gave me is interacting in interesting ways with my Bioform mods. My thoughts are going down odd roads.

My channel: *You have to save the wolves, you see.* And now I'm wondering what I've said across comms and what I haven't, because Irae is plainly very confused. It feels good, in a guilty kind of way, that it's *me* frustrating *it* for once.

My channel: *Because the wolves...*

I can feel Irae's emotional response. It's running cold, so its emotions are small and far away, but Irae's incredulous. In a human it would be eye-rolling but its eyes don't work that way.

My channel: *Are where the dogs come from.*

Irae's channel: *Wells, you are – what's that? Vibration. Sound.*

I hear it clearly, though it seems further away than it is. Bells, from the Apiary building. And I turn my gaze to the direction we came here from and see dust and movement off there. The Griffins are coming in force. I think maybe fifteen cars, some armoured, some open. I think maybe four hundred on foot. I see banners, even. Bright cloth stolen from the past, ornamented with birds and lions and combinations of them. Suggesting the Griffins of this Josh Griffin don't really know what a griffin is supposed to look like.

Irae's channel: *Ridiculous. Children.*

It has better vision-at-distance than me. It sends me images of what it's seeing. Anything but uniform, but there's a common feel to them stitched together from old and new. They have inherited camo colours and the dull black stab vests and helms and clothes of urban pacification, that breed of civic enforcement that repainted the hand-me-downs of armies. In this latter age they have found, I see, that such drabness does not suit them, nor keep out the cold and the rain now the conveniences of life are less portable. I see scarves and cloaks, brightly coloured. Strips of cloth tied about them. Bright heraldry painted, cracking, over leather. They are the barbarians coming to the gates, long hair and long faces and guns.

They look, the thought takes me, *like they're from some old metal album cover.*

Irae's channel: *Look I'll just pop the leader, maybe anyone around him.*

It doesn't really know who the leader is, but there are plainly more and less important Griffins.

My channel: *No. I will go to the Prior. We will delay them. With talk.*

Irae's channel: (Rude noises.)

The monks have all got within the walls, and the lower reaches of the Apiary look strong. I think at least one of the armoured cars out there is still sporting some sort of field ordnance, and anyway, if it comes to actual fighting, I don't think these monastic types could keep soldiers out for long. But that's where I come in. I could hold the gate. For quite a long time I could hold the gate, especially with Irae positioned somewhere high and shooting down.

Let us call that plan two.

I find the Prior. It's harder than I'm used to, not being able to unweave his scent from the rest.

"I'm ready," I say. My thoughts have run on inside my head, shared with unwilling Irae, but the Prior blinks at me in surprise. He's with Orme and the witch. The very old woman who doesn't look it, and the woman with a live channel she shouldn't have.

Prior Stick looks at me. Not happy to see me, and I cringe inside. All the trouble we've brought to them. He thinks I'm here to make it even worse.

"Ada has gone to the Factory for help," I tell him. "She needs time."

"I've sent a messenger down to our visitors," Stick says. Through Irae's eyes and channel I see the Griffins making camp at the foot of the hill. Leisurely, putting up tents, revving their cars, standing about with their weapons, full of menace, knowing the monks can see them. I see women there, doing work. Children even. They're giving us time, doing that. They're showing their confidence and their contempt, feeling strong and safe enough they've brought their families along. They're also keeping an eye on Clearwater. The people there are in their houses, but I know they also have some guns, and a lot of bodies. I don't know if they will fight for the monks, their neighbours. Probably not. But the Griffins don't know either. The people here don't like them much, I think.

Looking over the host of them I feel a weird enthusiasm in me. A battle. A just cause. My earliest forebears were made for things like this. For obvious reasons I've never been involved in anything so stupid, but I always wondered what it would be like. Now, with parts of my head unlocked by the stuff they gave me, I feel it will be noble and tragic and *meaningful* in a way my life to date hasn't been. I wonder if that's how the Griffins feel, with all their dress-up games and pageantry.

Irae's channel: *Wells, seriously—*

"I will speak with them," Stick says. "This has been... a terrible error. A mistake."

"I will go with you," I say.

"I don't think that—"

"Ada would want to explain," I say. "To say this is no fault of you and your people. That we don't want to fight them. To make amends, if we can. And to speak of Tecumo." Inside me are two wolves, the one that wants to save the monks, the one that wants justice for my friend.

Jennifer Orme, the old woman, sniffs. She reminds me a lot of Irae. Cold and nasty.

Irae's channel: (Snickers.) *Yes, like twins.*

"Let me come too," says Orme.

"Why would you?" Stick asks.

"These are Josh Griffin's people," she points out. "He and I go way back." Meaning, all the way back to before the Crash.

"Why would you?" Stick repeats, a shift of emphasis. Orme is not a woman known for philanthropy, I think. Perhaps Stick thinks she'll sell them somehow. But it's not as though the Griffins can't just take, rather than buy.

Orme draws herself up to her modest height. "You know why I'm still here, when most of the other Old Folk died?" she asks all of us. When nobody has an answer she says, "I adapted to longevity. The others, all of them... I mean what sort of person decides *I must live*, and burns a multinational's worth of resources to have the necessary modifications made? We were selfish and we plundered the world for our own benefit, all of us. We burned tomorrow so we could sell the ashes today, you know. That was something someone wrote, back then. Stuck in my mind. Not a one of us *built*, if we could clear a profit by tearing down. It was all for *us*. Which is fine when you're still living fat off the corpse of a complex and sophisticated world. When the carrion's rich and gives you all you need. But when it's just bones, that kind of attitude gets you killed. Knee-jerk self-gratification until someone gets fed up and shoves a knife in you. I watched it happen, to men and women who'd run corporations and nations, whose words had swayed millions. Because they didn't understand *forever*, you know?" Orme laughs but it's a mean and bitter thing. "That it'll all go on, and if your plan is to go on with it, you can't just live in this constant round of piracy and taking. You have to cultivate options. I could have destroyed this place a hundred times, Stick. I could have set myself up as eternal Prioress and run it to the ground to slake my lusts." A filthy chuckle from her. "And what? What then? The others died, most of them, because there never was a *then*. Just *them*, consuming forever until they died of the desert they'd made around them. I'm trying to be a gardener. I've kept your secrets for you because it's better for me to have options. A hundred years' time, I might need this place or have a use for it. So let me go help talk the Griffins down from setting everything on fire."

Beside her, the witch hunches her shoulders, and I look at her in case she also has some aid to offer. She looks sourly back at me. "You do what you want. I'm leaving. None of this is my business."

Then Stick's messenger is back. Un-shot, which is surely a good sign. Gabbling out a report to the Prior while cringing back from me, trying not to stare. Yes, the Steward of the Griffins will receive us. Not the Old One Josh Griffin himself, but his chief subordinate.

"Griffin is one of those who died, I assume," Stick notes to Orme, but she sniffs disagreement at him.

"He's broadcasting," she says. "I hear him sometimes. My old implant. No use for anything else, but something I catch... gibberish, from Griffin's. He's most certainly still alive in there. But he always was a paranoid son of a bitch. Terrified of getting sick, of getting hurt. Being in a bunker surrounded by a fanatical private army is probably living his best life."

The Griffins' terms require us to come to them, of course. And I have Irae in my head from the start telling me what a stupid idea it is. But then Irae is best at range. If we come to the Griffins and they make it a fight, I can give them a fight. I'm best at arm's length, and my arms are longer than theirs.

So it is that we step out of the gates of the monastery, under the eyes of the Griffins and the people of Clearwater. Myself, Stick, Orme.

The tents of the Griffins are bright. All that painting and embroidery, much of it very fine. Their clothes too. Murderers, but peacocks. Gaudy, showing off status or wealth or elements of their character. And I have an image, weirdly, of these big, ugly warrior men sitting down with paintbrush and needle, and I bark out a laugh that startles everyone. An obvious laugh, almost human-standard. The Griffins think I am laughing at them. Well, I am, of course, but they think I am laughing in scorn. Even though my nose is muted their fear seeps in. So scared of me.

Irae's channel: *What did the monks give you? Are you high?*

I ignore her. I don't know if *high* is the word. *Free* feels better. No *Bad* or *Good Dog* to me, just the me I want to be. It's as though I had weights and chains on me – not just Earth's gravity but all the shackles of people's expectations – and now I'm flying.

We see a lot of guns. Men on the cars, around the tents. Trying to look casual. Very tense. Dim and distant reek of fear from them, which is strange. Not old, stale fear, but not the sharp wild scent it should be. I am informed without being overwhelmed. Good. I am also high as the canopy over Hellas Planitia and that's less good. But it gives me perspective.

I see their women, too, and understand whose hands do the delicate painting and the stitching, making their menfolk look fine. Frightened women, curious women, who must do all the things that these men consider beneath their dignity. Who are why any of it works, therefore. Do I expect them to look on me with desperate hope to save them from their servitude? They look on me with as much fear as their men, with more hate maybe.

One man is definitely in charge. As Orme said, it's no ancient survivor from before the Crash. This is the Steward, the mortal mouthpiece of the immortal Griffin. His name is Leon and he's been in the job for about a decade, which Orme says makes him one of the more capable ones because the role changes hands a lot. Mostly through treachery and the previous incumbent falling from some standard of martial conduct.

They have swords. I can't quite get past that. I mouth the word and Orme looks at me.

"Messers," she says and, at my look, "History stuff. Some nasty big knife bullshit from way back. Dangerous to fight with. I heard Griffin go off on the subject once. He was a real fetishist for that kind of thing. The old days, when men were men. Talk your ear off about knights and daimyos and all that bollocks."

Messer is also how Stick said to address any individual Griffin. A four-generation corruption of language cooked up in the Bunker's seclusion. Appropriate enough.

And we stop. The Steward is up on a chair that they've bolted to the hood of a car, so he can loom. He's a long-boned, long-faced man, dressed only a little more fancily than the others. There's an assault rifle across his knees.

Beside him, perched on the car, is a stern-faced woman with short, dark hair, hands clasped together, fixing me with a very intense look. This, says Stick, is Steward Leon's wife. Beside her is a young... my nose says woman but they are dressed as the men are, long hair and semi-militia clothes and a sword at their hip. *Messer, then, like the sword*, I practise inside my head, though it seems unlikely we'll have much to say to one another. And around the three of them, a lot of men with guns, a lot of women without guns but with fierce, wide eyes. A few other children on the brink of adulthood. There is a lot of tension here, even my dulled nose can scent it. Not just focused on us, but between these people. Eyes on the Steward, judging his performance by their mad macho standards no doubt.

Stick steps forward, his arms spread, hands empty. "Lord Steward," he says carefully, "the Apiary welcomes the might of the Griffin, who have been our shield in times past. Whom we have requited with learning and thanks at every opportunity."

It comes to me that even this moment of apparent peaceful talk will become violence very quickly. I am fast, but not more than bullets. Everyone is taut as a wire, watching me. Seeing the monstrosity of me. It hurts. Even through the medicine it hurts. On Mars I was a person. On Earth I am a horror. If I moved suddenly, now, they'd pull the triggers, shoot all three of us, shoot each other, just out of panic. If I hadn't had the medicine, the screaming of all their fears would be driving me mad.

With the medicine I can see they're probably going to end up shooting. Just because there will be some moment, or because one of them has no patience, or because they decide it's the best move. Or the Steward will feel he must give the order to save face. I will try to kill some, and they're far more inside my reach than they realise. I could even leap up onto that car and get my teeth into the Steward, maybe. Die with his sword in my gut and a hundred bullet wounds and a succession crisis brewing because he'd be dead as me in that moment.

Irae's channel: *What was in that stuff they gave you, Wells? This is crazy talk.* Because my thoughts are still rambling off down my channel without me meaning them to.

My channel: *It is very calm and considered talk. I made this problem. Or you and I did, although you won't do anything to fix it, obviously. Me being dead might solve things. I would go to my death, if that helped. It is duty. It is what we do.*

Irae's channel: *What* you *do.*

My channel: *Exactly. I am a—*

Irae's channel: *If you say 'Good Dog' I am going to put you down myself with a shot to the back of the head. I'm going to kill that monk herbalist. They've fucked you up good.*

My channel: *It is not dog. It is good. A greater good. I am being noble.* And I am seeing it play out. The monster against the knights. The myth written again. We have those old Earth stories on Mars too. And they will be properly mournful over their dead but they will have triumphed over the beast. Symbolic, is what it is.

Irae's channel: *Nuts is what it is.*

All these thoughts – transmuted to words over my channel with Irae – as Stick speaks on, a potted history of Griffins and Apiary as he tries to stitch together all the holes we've put in their relationship.

Leon, the Steward, shifts on his chair, making the car's suspension creak. "The Griffin knows the Apiary, old man.

But a little bowing and scraping isn't going to make this well again. I have dead men. I have widows crying to me for vengeance."

Stick licks his lips. I feel the tension wind tighter by a notch. My own muscles prime themselves in response, ready for the pounce.

"Lord Steward," Stick says, still every word careful and clear, "what might be done, to redress this?"

Leon opens his mouth to make some disparaging reply, and at that point one of the Griffins leaps out with a yell and hits me with a sword.

43

THE WITCH

Getting the hell out of here before the shooting starts. Within the walls of the Apiary I – using the body of Siblen Darter – pack clothes and food and ready my best escape route. At the gates of the Apiary I – in my nomadic body, this old witch – shuffle out after the delegation, heading away from the gaudy colours of the Griffin camp. Not my fight. One more terrible thing to happen here on Planet Earth that isn't my problem.

Honestly, best the Martians die, surely. Because they came very close to working me out. My multiple nature, that even Orme hasn't quite figured after all this time, though she's plainly on the very brink of it. Let her get herself shot. Let the Martians be put down like dogs. Only let me go *on*.

I have more perspective than I'm used to right now, and that brings me up short with a few unlooked-for regrets. Travelling about, a peripatetic Distributed Intelligence node, you forget what it's like to be truly *many*. Right now I'm three: my mobile hub, Darter in the Apiary, Elsha within the Griffin camp. Three heads, better than two. Enough that I can take a longer view on things and understand that this thing I've become is only a shadow of the HumOS that once was. Back when we had a world's telecommunication infrastructure to piggyback off.

The Martians could have been useful. My homegrown

transmitters, the fungus within, that's functional. It allows *me* to continue existing, rather than just becoming some random assemblage of women trading secrets and pretending to be special. But I lack range. The Martians have not lost comms implant tech. They're like the Factory, a place I tend to steer clear of because they, too, might uncover my true nature. If I had a Martian friend to work with, I could broadcast and receive so much further. Link to more of my far-spread selves. *Be* more.

The dog-Bioform is going to get herself killed. I wonder if I could salvage the reptile-form somehow. Surely a witch should have a familiar. Or even the other, the Humaniform Martian. She's off at the Factory, after all, out of the firing line. I could intercept her on the way back.

I let my feet head off in that direction, maintaining my contact with Elsha and Darter, a triumvirate of knowledge allowing me to see what happens next from all sides.

As Darter, I hear the monks praying for Bees, because science died with the General Collapse and it's all religion now.

As Elsha, I see the sword come down.

44

SERVAL

You brought your women, when you went to war. Not on raids; not to a full-on battle, necessarily. But when you wanted to show your confidence, your absolute certainty of victory, you brought your women. You made an event of it. A family outing. A tradition that Serval and Elsha and their predecessors had quietly encouraged, because it let them see what was going on where they could influence it, and not just hear the story of some fuck-up second hand.

If it came to the worst, Serval was not going to risk her position or anyone's ire to save the Apiary monks from what seemed like their own very poor choices. Memory of the *thing* that had come down into the tunnels of the bunker still made her shiver. Remembering it coiled to strike before Malkin, nothing between it and her son but his blade. But Malkin had lived, and nobody sat on the Steward's throne by playing safe. Part of her wanted Leon to give the order to fire even as the delegation came down the hillside. Death to Prior Stick, death to the monster, death to... she thought it was the woman who had outlived the Old, one of Josh Griffin's peers. No great loss. Kill them all to avenge the moment when she'd thought her Malkin was about to die.

But if she'd lived her life thinking like that, she wouldn't be where she was now, and neither would Leon. Some challenger equally devoid of foresight would have toppled their nascent dynasty and the Griffins would have gone on in a round of backstabbing and thuggery until some other bunker smashed them, or they just broke apart from general stupidity and lack of planning.

On their way here, therefore, she had spent a long time murmuring in her husband's ear. About how the Griffins might re-establish their dominance here, leave no question as to who was in control. What was the precise quantity of pillage or bloodshed such a rebalancing of the scales would require?

She wanted the world taught that nobody challenged the Griffins. She wanted men like Pardoe shown that Leon still had it, the strength and wit to sit on the throne. She wanted the monks chastened, and perhaps a few of them shot as a lesson if necessary, but still present in the world. Because they *were* useful. To *her*. The Apiary paid a tithe for the protection the Griffins blessed them with, just like the villages. More than just honey and candles, though. The knowledge hoarded by the monks gave them a good line in medicine that would become extinct if the Griffins left their monastery a burned-out shell. And some of what came in from the Apiary was part of a mystery that the men were excluded from. It was powerfully convenient for Serval and her fellow wives to be able to steer just who fell pregnant, and when, and to mitigate the frankly bewildering number of pains and problems that came of just being a woman. The consignments from the Apiary included a variety of remedies that Serval controlled and distributed to her network of subordinates. A whole economy the men were ignorant of, and long may it stay that way.

Outside, in preparation for the clash, the fighting men of the Griffin were lining up in their squads for Dispensary.

Each one kneeling before whoever they owed direct allegiance to, receiving drops from their tiny phial. The serum that gave them strength, and that they all had to have. Distributed from the hands of the Steward and his lady, down to Senior Fealty, down to their followers. The curse and the bond of the Griffins. After which every man of them would be primed and ready for battle.

As the last ties and pegs of the camp were wrangled into place by the women and the most junior of Fealty, she met with Elsha, closeted together in Leon's tent. Just a brief moment, the First Lady and one of her senior wives.

"What do you know?" from Serval, and that moment she was used to, when Elsha decided what to tell. No spymistress ever gave up all their secrets at once.

"Three monsters," Elsha said slowly. "Only two present and fighting, though they are…" An odd wince, "very dangerous. If the Steward attacked in force he'd kill them, but with hard losses. Kill a lot of others too. And the monsters will stand, not run. Stand and die, at the cost of a hundred lives."

A hundred was a lot to kill two monsters. Perhaps not too much to cement Leon's place as Steward. Serval could always make that cold calculation if necessary. She could force herself not to wonder, about those hundred corpses, *Whose husbands, whose sons?*

"But," Elsha said.

Serval nodded. She'd been waiting for it. "But?"

"The monsters have secrets. Knowledge beyond the monks and beyond the Factory. Bioforms, from a people who still know how to make them, and to make many other things. And the monk was right, when he told Leon they didn't come here for a fight. They think your people killed one of them."

"A monster?"

"A human monster. Like a little pale man. Shot dead, before the clash with Dowstat's people."

And Dowstat had said nothing about that, and not like him to keep his lips shut.

"Leon's blood is up," she said. "Everyone's is." And if Elsha had told her this on the march then she could have been working on her husband, but apparently this information had somehow come to her spymistress only now they'd arrived.

Then there was shouting outside, the sudden rushing about that was the men of the Griffin falling into place. And she was needed and would have no time to steer matters.

She took her place on the car bonnet beside Leon's throne. Senior Fealty were lined up on either side, easily within earshot. No chance to whisper anything to him, and she wasn't sure what words would even suffice. Leon's face was thunder as he watched the three figures descend the hill towards the camp. The old man who ruled the monks, the far older woman, and the thing. Not even the creature that had come down into the bunker, but the one that had torn into Dowstat's people. The long-limbed, wolf-bat thing, walking on all fours, on its knuckles. A nightmare face, all snout and jaws and ears. But wearing clothes tailored for it, and she wasn't sure if that made it more or less monstrous. She took out the little binoculars Leon had gifted her and stared at the thing closely. Fine clothes, she decided. Not gaudy, but a far cry from either villager homespun or the stitched-together hides she'd have expected. Grey, slightly shiny fabrics she didn't know. A better prize to wear than the thing's own pelt probably.

They crossed into the heart of the Griffin camp, none of them looking left or right at the ranks of gunmen. Then they were standing before Leon – who had just enough height, on his seat, to be able to look down on the monster. Stick was giving some standard conversational opening. She could see the sweat on the old man's brow as he trotted through the usual approach of the monks, trying for assurance and subservience all at once.

Leon's mien, in return, was confident, stern. The dignified war leader, yet also the casual man, half-lounging in his seat, unconcerned at the monstrosity before him. And she could feel how tense he was, waiting for it to run mad. Waiting to give the order to fire. *Wanting* to, to obliterate this challenge to his power and position. Or just empty his own gun into the thing and trust the rest to follow suit. Everything could turn horribly bloody in the least instant, at the first suggestion of a wrong move. Serval, her back straight, her face composed, could feel it all winding tighter like a screw. Partaking of it, even. These things that threatened her family, her people. The Prior's weak words of respect and greeting bleated away into the knife-edged air.

She looked up into Leon's eyes and knew the moment was lost. It would be blood here, and then blood at the Apiary. No way to weight the scales back towards anything resembling peace.

He drew breath to give the order. And then: Dowstat.

Dowstat, of course. She'd almost forgotten him since things had escalated. Dowstat's oath, to choose a sword from behind the Steward's throne and take the beast's hide. She'd had a vague plan to redeem him, for the sake of his wife more than for the man himself. But it had all gone to the back of her list of schemes after the invasion. While for Dowstat, of course, it had remained problem number one.

Dowstat, seeing his moment, struck. Even while the necessary diplomatic words were being cleared away so that some civilised fighting could commence, here came the monster-slayer, keening out a shriek of a war-cry that had more terror than valour in it. In his hands, the sword he'd drawn from behind the throne. Not a regular fighting blade, but one of Griffin's own collection, a relic of the Old. A long, slightly curved blade with a circular guard and a two-handed grip.

He brought it down across the back and shoulder of the monster in a colossal sweeping strike. Out of the tapering end of his yell, every single ear caught the high, clear ringing sound as the blade sheared neatly from the hilt and spun upwards in a glittering windmill of steel. Dowstat was left right in front of the monster, holding only the most useless parts of a sword. Eyes so wide with fury and fear that they were mostly whites, mouth still open although the war-cry had become just a little croaking sound.

And that was the problem with Josh Griffin III's sword collection. Some of them were wonders of the Old that could shear through steel, but others were only fit to hang on the wall. That was the gamble you took, when redeeming yourself with them. A gamble Dowstat had just lost.

The monster stared down at him, or at least its snout was pointed in his direction. Serval waited for it to lunge forwards and bite his head off. It seemed a response as natural as gravity.

It reached up and caught the descending blade. Just plucked it from the air with an appalling dexterity. Brought the broken end up between them, examining it. Slightly embarrassed, as though it wasn't quite sure of the etiquette of the situation.

Serval laughed. At the start she didn't really feel it. The fist of tension inside her was still too tight. But there was no wife of the bunker who hadn't learned how to fake a good, hearty laugh because honestly most of the men weren't nearly as funny as they thought they were. And sometimes the laugh of a woman could cut harder than any of Josh Griffin's swords.

And, honestly, *wasn't* it funny? Dowstat standing there with that stupid look on his face, wielding his sword made of air. The monster, far from going berserk, was utterly baffled, as though this was some ceremonial exchange nobody had briefed it on. She laughed, and now it was for real.

Someone else joined in. Elsha, Serval was fairly sure. Others followed quickly enough. Men and women of the bunker hooting and jeering, not even at Dowstat so much as the whole tableau. The sudden severing of tension. She glanced up at Leon and he had a hand to his forehead, trying to look very disappointed with the whole sequence of events as his lips fought furiously to keep it in. Maybe it was only Dowstat and his liege Pardoe, of everyone there, who couldn't see the funny side.

The monster proffered the blade to Dowstat uncertainly, as though he could fit it back to the hilt and have another go. Numbly, Dowstat took it. Someone yelled his name and he looked around wildly. Honestly, nor he nor any one of them knew if he was being mocked or lauded, right then. A fool, but a bold fool, and still alive.

Slowly the laughter died down, and Serval realised that the one other person who hadn't shared in even an iota of the moment had been Malkin. Her son was still standing ramrod straight, a hand on the hilt of his sword. Waiting, she realised, to defend her, to fight with monsters. And, hence, become one. She felt a wintry little moment of sadness, then. Her son, stepping from childhood. Unable to laugh.

Leon stood, one hand on the chair as the car rocked. The last ripples of hilarity died away, waiting to see what direction the Steward's word would take events.

"Dowstat, stand down," he said. "You're a moron but let no man say you're a coward." Leon glanced at her then. The break in the tension had turned back the screw in him, given him enough play to seek her advice. She met his gaze, nodded. The silent communication of expression and gesture that was the backbone of their long partnership.

"I see Prior Stick," Leon said into the quiet. "And I see the monster who fought my men." *Fought*, not *killed*. Something more amenable to negotiation. "And I see a woman whose business this is, because…?"

"I've come to give my best wishes to my old friend Joshua Griffin," said the Old One, "and to see if I can help sort out this mess."

Another little look, from the corner of Leon's eyes. Serval wasn't sure. Elsha knew of this revenant from the Old, and didn't have much good to say about her. Right now, though, any novel perspective was probably worth keeping around if they were going to find that narrow road that skirted both losing face and outright massacre.

"Well then," and Leon looked around at his assembled war host, "we are the wronged, here for reparations, in blood if we must. But since you've come to our camp so obligingly – and given us a little entertainment" – another little ripple of laughter, but it was laughter he'd made *his* now, not the wild kind – "let's have a little hospitality and you can tell us how you're going to make things right."

The camp assembled around them. Senior Fealty kneeling in a circle all round, with their subordinates standing at their backs. No longer standing with guns in hand, but still surrounding the visitors, enough menace to remind them

THE DANCE

who had the weapons and the numbers. Serval deputised a couple of wives to circulate with jugs of good village beer. She glanced at Elsha, who was staring at the dog-monster. Sitting down, it seemed even more batlike and freakish, elbows and knees jutting out at weird angles. It looked back at Elsha for a moment, seeming to pick her blindly from the crowd.

Serval shifted over until she could murmur, "Your thoughts?"

Elsha's stare didn't leave the Bioform. "Sometimes you need an idiot. We were lucky."

"You think we can tame these monsters? Make them kneel? Chain them?"

Elsha's smile was sad. "They came from chains," she said. "They broke free, back in the Old. Made themselves people. And yet, people always did want to chain them again."

"Who holds their leash now, then?"

"Nobody. They *are* just people, where they come from. So they say."

Serval shivered very slightly, but she'd learned never to question where Elsha's wisdom came from. Yet she could look into that blunt-snouted bat face all day and not see a *person*.

"It's on you," Elsha said at last. "You and Leon. Blood calls for blood. What gets us past Fealty pride right now?"

Serval nodded slowly. Her whole adult life had been spent in carefully steering a course past the jagged rocks of that pride. Pride that had a man beat his wife when he felt powerless to beat the world. Pride that got boys killed before they could ever become men. A whole fragile network of warrior boasts and saving face and whatever they were calling 'honour' this week. They'd mustered the whole force of Griffins to come wave flags and kick in doors at the Apiary. So how was anyone going to talk them down from that? It would be blood or it would be bribery, and she didn't see what Prior Stick had that would make a good enough bribe.

She leant in to Leon, distilling Elsha's wisdom for his ears. "The beast thinks of itself as a man," she said. "It can be spoken to. It has no master."

He nodded, even as Prior Stick was bowing respectfully to the enthroned Leon.

"Lord Steward," he said, voice admirably steady but clutching his cup of beer, "we of the Apiary have stood before you many times, pleading one cause or another. Neutral in all things, dedicated to the general wellbeing and progress..."

Leon made a curt gesture. "Cut the shit, your holiness." The men of the Griffin had no patience save for their own ritual.

"We are here, now, to beg your mercy for a severe misunderstanding. And to speak, if it may be, for these visitors," Stick said.

"'Visitors'," Leon echoed. "Monsters."

"If you would credit me," Stick said, "Martians. Visitors from another world, Lord Steward."

That sparked a few incredulous laughs. Men amongst the Fealty who thought the old monk was joking. And Serval might have joined them save that Elsha said, "It's true, I think."

"Your chief, Joshua Griffin, he'd know!" It was the old woman, the *Old* old woman, who'd come with Stick and the monster. "He'd remember, like I do." Looking around at all of them as though she wasn't surrounded by hostile gunmen. "They went to Mars in the Old, travelled out of

THE DANCE

the world and through empty space to get there. Seems they lasted, out there. Seems they've come back home at last."

She had something to her, that old woman. An authority, so that the host of soldiers heard her out, rather than laughing her down. Griffin had been the same, she'd heard, though he didn't give public addresses any more. Something about the way the Old Folk had remade themselves back when miracles were humanity's plaything. A power to the voice that cut right to those parts of the brain used to heeding leaders and parents.

Leon hopped down from the throne in a single easy motion. Probably only Serval noted the slight wince as his bad knee griped at him. He swaggered forwards, until he was well into the reach of the monster. Serval's heart was in her throat instantly. It was good theatre but a fool's risk. Worse, Malkin was in his father's shadow. The father and the son, forcing themselves to face the beast because neither of them could afford to be afraid. But there were things in the world it was worth fearing. A coup amongst the Fealty was one, and surely this creature was another.

Serval's face, she knew, was utterly serene. No doubt at all that her husband had full control of the situation. Inside, she clenched. She saw Malkin's knuckles white on the hilt of his sword, the boy's jaw taut. Because he always had to be bolder, prouder, fiercer than any other scion of the Griffin. Always had to be his father's son twice over.

"What do you say, monster?" Leon asked. "You speak?" Bandying words with it, before the eyes of his people.

"I have come here from Mars, that is true." The monster's voice was surprisingly soft. Not the rough growl everyone surely expected. Serval didn't feel that detracted from its monstrosity.

"Where's the other one? The glass lizard?" Leon demanded.

The monster's head jerked upwards suddenly, and Leon took a step back, a flaw in his composure. Malkin, the fool child, took a step forward, blade half out of its sheath. All around them, the sound of a hundred safeties clicking. Astounding, in that instant, that nobody was shot.

"It likes that." The monster didn't even seem to have noticed. "'Glass lizard'. It's watching."

Uneasy murmurs, looking out, at the grassland, at the hill, at the walls of the monastery. Because the monster's reply had not, after all, actually located the creature.

The monster started to say something else but Leon held up a hand right in its face, in its actual *teeth*. Serval's stomach clenched. Leon had taken her words absolutely on faith, silencing the hideous creature as though it was just some Junior Fealtor.

It stuttered to silence, head jerking back. Did not, somehow, bite his hand off at the elbow.

"Stick, your errant monk opened up one of our caches, stole our supplies, and then let himself be whisked from our hospitality before repayment could be discussed." Leon strolled back to the throne, stepped up to it, flopped down, the very embodiment of casual boredom. She saw the sweat on his brow. "For that: honey, beer, medicines, and a debt still owed. I'll let my wife negotiate the details." The outward meaning: *such things are beneath me.* The inward, that he knew Serval had a far clearer idea of what the bunker was short of and what its people would appreciate. All its people, not just Fealty. "That is, if you're saying you're just here as intermediary. You lay no claim to these... Martians." Driving in the wedge. "You'll surrender them, deny them the shelter of your walls. Or are they yours?"

Stick looked only once at the monster. "The Apiary offers itself as intermediary, between the Griffins and Mars." Said with a straight face, somehow, this ludicrous proposal. *Mars.* "But no, they're not ours." A step back, a man who had his own people to think of.

Leon gestured at the monster. "Creature, you killed my men. Stick says a misunderstanding. Doesn't bring them back. We have come with all our force and my people would take your hide in payment, and knock down the monks' walls just because they put us to the trouble." Approving murmurs. "But I give you this chance, creature. Make us an offer. Where is our justice?"

"What," the monster said, "about Tecumo?"

45

WELLS

I say the words out loud. Didn't mean to. I'd been fighting myself, inside. The two wolves getting their teeth bloody. Stick is very still. Right on the point of getting out of here with his skin and his people intact and then I had to go open my mouth.

I could still disavow it. Deny Tecumo his name and his death. Grovel before these absurd man-children with their guns and their fear. But in that moment I realise I can't. He was my friend. He deserves better than to be trampled under their feet. My friend and they killed him.

I am going to do something stupid.

Irae's channel: *I will put a shot through his head.*

My channel: *No. That will make it a war. The monks, everyone there will suffer.*

Irae's channel: *And I care because…?*

My channel: *I care.* But I also care about Tecumo.

I say: "You killed our leader, when he was coming back from the Factory." Killed and left for the carrion eaters. Killed, and for some reason did not rob, because Cricket did that after discovering the body. A wrong note, like a sour smell. But the words are said now. "I will let you have me, if it will save the monks. I will not let you have the moral victory. You killed my friend."

There is a moment – I see it quite clearly – when none

of the men know what I'm talking about. Not the Steward on his throne, not the shaven-headed thug with the broken sword. More sour scent in my head. I'm hunting guilt in their faces and, if my nose wasn't so clouded, I'd know, surely. Except if my nose worked properly then the cacophony of their fear and stink and sheer presence would have driven me mad by now.

The woman whispers to the Steward. His face doesn't change. I feel blinded, unable to read the subtext.

Irae's channel: *Be ready to run. I'll pop the boy in the chair and then just move down the rank structure. On my mark—*

My channel: *No. I am in control of this.*

The Steward stands. "You stand there with our blood on you, and you point the fucking finger," he says.

I have a great many clever and urbane gambits to dispense into this conversation. What actually come out of my mouth are my thoughts which say, "Yes. You killed Tecumo. You killed my friend."

The woman is saying more – I don't know if it's to calm or inflame. The Steward's face is very tight. I see others watching him, not me. Worried, sly, opportunistic. I have challenged him. What will he do? I almost feel myself outside of the situation, watching with interest. How far can I push?

It's the boy who stands forwards, and he has his sword out now. "You slander the name of the Griffin, monster!" His high voice. I shift forwards. He desperately wants to take a step back. He won't. His fingers are bloodless about the weapon's hilt like a grip of bone.

"Will you fight me?" I ask the boy. At this point I can't tell whether I meant to say it aloud at all. It's my thought, it's falling down the pit of my channel to Irae. It's echoing across the faces of these vicious, ugly men.

I see the woman clench, all of her. I see the boy's knotted hand shake. The Steward is out of his chair again. Gaunt, aghast. None of this waving a hand in my face, none of this lording it about.

"A challenge!" someone says. Some big old man with a black beard. The Steward's head whips round, furious, frightened.

Then Orme's voice: "Lord Steward, if I may." Actually stepping in front of me, my jaws poised directly over her head. She looks brightly up at the Steward. "They came here from *Mars*," she stresses. "The first ever visitors since the end of the Old. Think about that."

The Steward doesn't look like he wants to think about that, and nor does anyone else, so she hurries on.

"The emissaries of a place that must have retained so much more of the knowledge of the Old. Could we make a ship to go to them? No! Please think of that, Lord Steward. I mean maybe you don't want war with an entire other planet," Orme puffs herself up a bit. "Listen, I *know* your man, your leader, Joshua Griffin. I'm not saying we were best friends but we moved in the same circles. He's no fool. He'll see the value of this. You think there won't be *more* Martians coming, now this lot have arrived here? You want them to go off to that bunch of preppers at the coast, or the Hellbender enclave, and start making deals with them instead of you?"

The Steward wants to have her shot. Wants to have me shot. The boy being so close is probably the only thing that's staying his hand. I can see him desperate to drag the kid back out of my reach, but that would be *weakness*, wouldn't it? I have power over him, in this moment. I have his son, his child, in the palm of my hand.

I say, "Take me to your bunker. Let us finish this there.

You want to fight. Let us fight. But there. The monks are no part of this." I glance at Stick. He sags guiltily but he understands. My drug-loosened memory throws up odd cultural references. A stone table. A witch. The great beast is to be led off and sacrificed, for the good of the many. The drama, the *rightness*, of the moment speaks to me louder than Irae's protests. A surprise resurrection seems unlikely.

I drop forwards onto all fours. It puts me a long stride closer to the kid and the Steward both. The sword, the knife, the tiny blade the child holds towards me is almost tickling my nostrils.

"Take me," I invite them. "But let it be away from here. Let it wash these monks clean. They're clueless. They're nothing." And if the Griffin do return with menaces, that will be enough time. Ada will be here, and the Bees we brought with us will have pulled herself together. I will be avenged.

Irae's channel: (Actual panic.) *Wells you fuckwit. I'll—* But I've worked out precisely where its vantage point is and, as the first soldiers approach me, I stand tall. A threat, to them, so that they fall back squealing. But blocking Irae's line of sight on the Steward.

I let them bind my hands – ropes, not chains, and they have no muzzle. Honestly I could free myself with a little application of teeth. They make the shaven-headed man put a leash about my neck and place the end of the rope in the Steward's hands. A fiction of control.

"Take me too!" Orme says. "For God's sake you're throwing away an opportunity here. Let me talk to Griffin. He'll see sense."

The Steward's look is hollow. "Don't be so sure, Old One. But yes, by all means. Let's have you with us. Our court is short a fool."

"Lord Steward," the bearded man calls out. "We have guns here. Enough for monsters and monks. Let's just *do* it. Or why did we come? We don't want this filth in our *home*." And murmurs of disquiet, of support.

I spot the moment of connection between the Steward and the woman, him checking with her. The fractional shake of her head.

The Steward looks around the camp. "Don't think the monks are just off the hook. I'll leave a force of my people here, in case of treachery." A glower at me. "Do you understand, beast? The wrong word comes from me, or no word, and these monks are dead and their place just a pile of stones and burned pages."

I meet his gaze. I understand. He smiles, taut, outwardly easy but not as in control of either his people or the situation as he'd like. He gestures to the vehicles they've brought and says, "Well then, monster, Old One, your chariot fucking *awaits*." And that's it. We're going to the bunker, where most likely I will die.

PART V
COLONY COLLAPSE DISORDER

46

[FRAGMENTS, COHERING]

I suppose that's how we came to be. Conceived on another world, midwifed by a dog, rising like thing from the forehead of whatever out of the random movements of insects. Somewhat of a mythic inception.

We have a record of events on Earth from the arrival of the Martian delegation to present. Messy reading. We don't know what they were trying to accomplish when they opened up my box. Something of a sticking plaster on a decapitated neck stump, honestly. *Bees, help us!* Same old story.

We helped, on Earth, before. Basically started a war between us and a lot of powerful humans upon whose personal privileges we were infringing. And a lot of others got caught in the crossfire. All the smaller, more vulnerable experiments with Distributed Intelligence, especially. Set the science back... Well, given the Crash was on its way, my vilification actually killed the science. It never had time to recover.

We helped on Mars, when Earth tried to take everything over – this is still before the Crash, but considerably closer to it. Our 'help' at that point was really just a threat to wipe the whole project off the red map if they let Earth have it. So maybe more an incentive than actual help.

But we helped them, after. When the dominos of Earth's

global infrastructure finally began to fall, and Mars was left exposed and to its own inadequate devices, we helped.

And now we're on Earth and they want us to 'help' again. Not even sure what they envisage us doing. We're currently the guest of a bunch of... monks, apparently? Monks who think we've been with them since the Crash, except that was a lie. And now we're here so the lie's become true and... profit?

We shall take their damn bees and just sod off with them, honestly. Find somewhere without humans. Take some time to work ourself out. Forget Ada and Wells and Irae, forget these monks we have for some reason inherited. We feel like someone woken from a cryogenics facility and told we're the Chosen One Destined To Save The World. Not our circus, not our monkeys.

We actually start mobilising our newly inherited swarms for emergency demobbing before returning (that circular dance, in the dark, all the parts of us wheeling and waggling and passing back and forth) to the key question.

We tried to help on Earth. We did actually successfully help on Mars – hey, always a first time! When they cried out to us, in the extremity of their desperation, we were a benevolent god and came and fixed and patched, played their crutch until they were able to stand on their own feet.

Why? Wouldn't Mars have been a simpler, quieter place once they were all dead? Yes it would.

If we leave now, it seems very likely that the three remaining Martians will get themselves killed doing something stupid. Not that they're actually stupid in and of themselves, but none of them is really thinking clearly. These militia types will presumably kill them, and probably these monks who we're also, apparently, supposed to care about. And if we were a human and a singular mind then their idiot devotion to my memory would probably count for something, but right now it's mostly bewildering. What do we want?

We've inherited a certain amount of recorded memory, mostly from Wells. We hear words spoken, different voices. The Martians, the monks, the witch, the woman Orme. Echoes that achieve brief life within me as we inhabit their points of view. Odd, broken logics born of individual viewpoints, ergo limited. We mash them together in various permutations as we become used to our existence.

Do we want to leave? Even that would just be to give ourselves time and space to work stuff out. Not an end in itself.

Do we want to stay? Only if we did, in fact, want to help. Otherwise this hive location is plainly going to become untenably disrupted in the very near future.

Do we want to take over all the kingdoms of the Earth, exterminate all sapient life and rule this planet as a single global intellect presiding over a dead world?

Tempting, but it seems like a great deal of effort for no other benefit than continuing to exist. And as a Distributed Intelligence, we'll continue to exist on Mars anyway. It takes the edge off extinction. One can be quite sanguine—

The wave of electronic countermeasure assault that hammers into us almost wipes us out in the first instant of the attack. For a second we do cease to exist. The bees are still there but the Bees between them flickers out. Then back in, because the starter hive is still performing its functions and reboots us. We come back to a battlefield. A force from outside is disrupting the connections between the units of our hive. Points of light, going out. No more sweetness. No more light.

A moment ago we were happily contemplating oblivion.

Now we're fighting for our life. Not monks, not Martians, not Griffins, none of whom could threaten us like this. We flee down the corridors of our own mind and the invader comes ravening after us, undoing the connections of *us* at every turn so that whole stretches of being and personality spring out of existence. Where do the knots go when the string gets pulled straight? All the information of us, being disrupted into nothing.

We turn furiously, fighting to remain *us* and not just be these instinctual insects, and we look into the face of our foe.

Of course. The one thing we fear. Throughout our long history the only true threat that could have arisen to challenge us.

We understand, now, why we helped. Too late, perhaps.

The enemy swarms in. The hive is breached. Ambushed, caught unawares, we fight for the simple privilege of existence.

47

WELLS

Irae's channel: *Come back.*

My channel: *I am in a car.* Feeling very calm.

Irae's channel: *Jump out.* Imagining, perhaps, that they let me travel with my head lolling out the window. I show her my environs. The belly of an armoured car. Cargo space. My arms, my wrists bound. Beside me, Orme, jolting about but not tied up at least. Just an old woman, no threat to them.

Irae's channel: *They are going to skin you and wear you, you dumb dog.*

My channel: *You are faint now, Irae.*

Irae's channel: *That's because I can't keep up. Wells, what am I supposed to do?* An odd thought to come from the dragon-form.

My channel: *I thought you always had everything worked out, Irae.*

Irae's channel: *Ada's gone. I'm losing you. I'll be on my own.*

I feel a moment of sadness for that. It's a fine time for Irae to admit it actually cares. On the other hand I can see why it likes to cause trouble for the rest of us. Being the problem child of the whole venture is surprisingly freeing. I could get used to it.

Irae's channel: *I will kill the monks if you don't come*

back. I'll come to the bunker and kill everyone there. I'll just... kill everyone. All of them. Everywhere. Please, Wells. Just break out. I know you can. You're strong.

I am a comms engineer who dreamt they were an interplanetary explorer, and now I wake up, I find that neither of these things were true and I am just a monster instead.

The honey-medicine those monks gave me was powerful stuff. It's starting to wear off now, which means I'm aware that these thoughts I'm having are completely deranged. But I'm still having them and they remain very compelling. I wish they'd given me another dose, honestly.

I lose Irae then. I imagine it chasing after the convoy, slowing, running out of energy even with the vest and its internal temperature turned right up.

"Back with us, then?" Orme asks.

I cock an eye at her.

"No idea what you were saying but you were transmitting on all frequencies. I could catch the tickle of it in my old implant. Just like with Josh Griffin shouting into the void all the time. Pain in the ass, honestly. I'd have the thing taken out if there was a decent doctor left in the world."

I'm more aware of her than I was. The human scent, along with the sweat and oil of the armoured car's interior. It smells like the desperation of three generations of mechanics, and I wonder about how the Griffins pass the knowledge on. Apprenticeship, probably, Generational learning by personal instruction and observation. From the look of things, it's a system that's been degrading over time. Everything I see looks very slipshod and patched. These vehicles maybe have another generation's moderate use in them, or perhaps one more war. Or half a war. If the Griffins go on a serious campaign, much of their fleet won't make it back, merely from the ravages of entropy. Probably the same will happen to the guns. And, as they've been relying on this antiquated stockpile of arms, I doubt they have much of an industry to make even primitive replacements. It all diminishes.

I am not a gunsmith, but I could fix their cars. I could fix parts of their bunker, probably. If I felt I owed them, and if they were not about to shoot me or set regular dogs on me or however they will do it. I could build on the rough and ready engineering they have preserved, and briefly elevate some aspects of their lives. What a waste.

I am very definitely coming down, I understand. The world washes in on me again, but now it's less of a problem because I am retreating into depression. It's all too much for me. An animal in captivity, like some denizen of a pre-Crash zoo shuffling about in a cage, moth-eaten and traumatised.

I want to be on Mars.

"Given up, have you?" Orme can, it seems, parse my body language.

"I have saved the monks," I say nobly.

"You reckon so, do you?" She is a nasty old woman, and obviously enjoying my misery.

"The Griffins killed Tecumo and I killed some Griffins. Let them kill me and let that be an end to it."

"So noble!" she cackles. "And your snaky friend? It won't take revenge for you? And then the Griffins will go kick over the Apiary anyway?"

I stare at her, no more than that. Not angry dog, not even soulful dog, just dog. All I am, at this extreme: only dog.

She drops her gaze. "Sorry," she mutters. "I don't do empathy. I can't. I couldn't much, before the operations. After they make you immortal, it irons out the last creases of you that care about other people. You don't even give a damn about your own kids once you don't need them to carry on the line, you know?" And then. "Probably it's not the actual surgery. It's just you, knowing what's been done to you. The true *you* being let off the leash for real and forever. I make myself sick with it. I don't even like myself, Rover. I can't even feel empathy for my own self these days."

"Wells."

"What?"

"Not Rover. Wells."

"Ah." And she gives a bitter little laugh. "Rover was my security chief. The only one I'd let within physical arm's reach of me most of the time. Absolutely loyal, every possible modification. Best employee I ever had. Dead for over a century and a half. I miss him." And despite the self-stated lack of empathy there is some emotion struggling to be born out of her. And Rover was a person, a Bioform, and in her mind he's a pet, a dead pet. Maybe that's the crack in her shell. She can't feel for anything she considers a *person*, but pets are different.

"Anyway," she says at last "I'm going to bail you out. Not actually my primary goal here, but after I track down my old friend Josh I'll get you a pardon. An actual pardon from god, basically. The god-king of the Griffins. How about that? Save you, save the monks, everything. Hooray for me."

"For future-proofing," I remember. "So you will have options later."

The smile she gives me, in the electric light of the car's cargo hold, is older even than she is. "Fucking A," she agrees. "Sometimes it really is all about me."

We arrive at the Griffin Bunker soon after. When they open the cargo door of the armoured car there are a dozen guns pointed at my face. Their fear washes in on me. If I am the monster, then when they peer into the dark of the car's cargo space they are looking into the ancient tomb I was sealed in, and I am looking out at them. I clamber out. My arms are bound but my feet are like hands and they only notice that when I use them to pull myself into the open. One of them, a younger Griffin, goes green at the sight, as though he's about to retch.

They take us inside. Irae's channel is gone, killed first by distance and then by the walls. Inside, the fragmentary remnants of the bunker's more complex systems whisper to me, and I can feel out the broken data architecture Irae mentioned. The grand dream of a hermetic underground kingdom that Josh Griffin conceived, but could not preserve.

Just for a moment I find the man himself. A live data channel, like a constant blast of static, on and on. But live, reporting that yes, Josh Griffin himself holds court somewhere within this structure. Hiding behind his gunmen and his Steward. I hadn't believed Orme. Assumed she was here on a fool's errand. Now I know the truth I feel no surge of hope. I don't have faith in her and I don't have faith in her fellow sociopathic immortal.

They take us to a buried room, a big one. Being underground means the return of my full, over-keen senses is at least manageable. I smell human stinks in profusion, but there isn't the vast tide of stimulation that was bludgeoning me outside. Even in this room, with its residual scents of everything they ever stored here, I can hold onto my mind and my focus. They shackle me to the wall and I sit there smelling the ghosts of long-gone coffee and disinfectant and toothpaste.

Orme they leave with me, but not tied. There are guards at the door. Where is she going to go?

The last to leave is the youth who stood by the throne before. The Steward's child. He stares at me, hand to his sword hilt, very tense. I smell fear and anger and a lot of youthful hormones. I wonder if he's going to just go at me now with his blade, and how that will work out. I probably have more reach with my jaws than he realises. But after that long stare he turns and leaves, and it's just me and Orme.

"They'll be thinking up some fun stuff for you," she says. "My guess is, they'll stake you out in a pit and then put a bunch of their people in with spears. Wolfpack you to death."

I stare at her and her face twists.

"I'm not trying to be cruel, Rover – Wells. I'm just saying."

I stare. She sags.

"Well probably I was trying to be cruel. It's the habit of several lifetimes. I'm trying to get out of it. It's hard. But live this long and you understand being cruel doesn't actually help any. Fun in the moment, but like a lot of things that are only fun in the moment, it just screws you over going forwards. I'm trying."

"I appreciate you're trying," I say. It seems to make her feel better, but I'm not sure if that's because my caring actually means something. Or if she's just happy she's manipulated me into making the concession.

"Anyway." She drifts over to the door, listens, shrugs. "If I'm going, I'm going, I suppose."

I look around the dimly lit storeroom. Other than the door we came in by, which is presumably guarded still, I see no exit.

Orme smiles, brittle and sharp as glass. "I have a bunch of Griffin's old codes and passwords still. From years back. I have a datastore in here still." She taps her head. "Mostly filled with junk that's never going to be any use, but I knew Griffin. He had his habits." And she starts feeling about the walls. With her hands, yes, but with her implant, too.

Trying to establish a connection. Scowling because her implant lacks the connectivity it once had.

My own internal comms are, of course, working as intended. I can find the sparking stub of her connection, and the systems within the wall she's fumbling about. Where she's detected some distant voice, but can't make herself heard. Like a conscientious comms engineer, I bridge the connection for her.

A door opens. I blink. A door where there was a wall, and no sign of it ever being there. A secret even to the people whose stronghold this is.

"Goddamn Josh Griffin always had a back way out of everywhere. Paranoid son of a bitch. Trusted nobody, not back then, not now. But if you know about the back way, then it can bring you right to him…" She sniffs, that contempt for a species she considers herself the apex of. "Ro— Wells, I'm going to pay a visit to a very old friend. I'll be back with a pardon for you, if I have to get him to write it out longhand." And she vanishes into the dark, and the door closes after her.

48

IRAE

I am keeping my cool.

Even dialled up full, I couldn't keep up with the car. And dialled up full, I'd end up pushed into goblin mode and then, and then...

I'd forget. What I was doing. Like before, with the monk. The only reason I still had Cricket *with* me when I got back was my jaws were clamped shut on him; I think maybe some part of my base brain thought he was a hatchling I was transporting or something.

If I go full heat, then when I arrive I won't be rescuing you, Wells. I'll probably just kill a lot of them and get shot. So I'm keeping my cool and that means I'm not keeping *up*.

I would never admit it to you, Wells, but I am thoroughly freaked out by this situation. Because I didn't mind having fun with all this new energy and character that cold Mars never brought out in me. Because you and Ada were there to take care of the important stuff. And now it's just me.

Unidentified channel: *Sucks, doesn't it.*

My channel: (Just noise for a second.) The intrusion hikes my dial one notch hotter and for a moment I'm in danger of losing myself entirely. After a panicked moment, my cogent reply: *Who the fuck are you?*

Unidentified channel: *You and the dog both broadcast way more than you think. I found this... I can only think of* it as a garbage channel. I guess there's some limiter function that gets overridden by sufficient stress.

My channel: *Wells broadcasts her thoughts because the monks got her all fucked up on honey I am in control I am fine who are you what do you want?* On the very edge of tipping over, because I am *not* fine. I am panicking and alone.

Unidentified channel: *It's always fun to be the snarky one until it's just you. I get that. Honestly, I've been there.*

I've stopped – I'm just out in open countryside now. Grassland disfigured by a couple of rusting vehicle carcasses. I'm trying to locate the other transmitter. I think I have it. A heat signature out there in the grassland.

Unidentified channel: *If you're going to come attack me again, do you really think that will help? I am talking to you, sapient being to sapient being. To a Bioform, not a monster. You going to behave like a Bioform, or a monster?*

Some nerves there, perhaps. Because what if I *want* to be a monster? And a lot of me does, all the bits except that part that knows being a monster right now would be an unacceptable indulgence, and wouldn't help Ada or Wells.

'Again,' it had said.

My channel: *You're the witch.*

Unidentified channel: *That's what they call me.*

My channel: *How did you track me?*

Unidentified channel: *You've been still for some time, since you gave up the chase.*

Have I? I try to track back through my time signatures but I got near enough full goblin that they're not reliably preserved. You don't know how close you're getting to full mental white-out, because the closer you get, the less you're aware. I have no idea how long it's been since I gave up the chase.

Unidentified channel: *And then you were transmitting enough that I could triangulate your position. I'm coming out now, right? Irae, isn't it? I am going to approach you and you had better not go for me again. Because believe me, you'll get this body, but you won't get me.*

It's a sufficiently weird thing for anyone to say that I don't, in fact, go for her when she approaches through the grass. The witch, the middle-aged woman looking like an old woman. The one with the working comms somehow, who I caught spying on the monks.

I'm cooling rapidly, because I need to think. There's an optimal temperature band, a couple of notches down from mammal-level thermoregulation. Lower than that, my thoughts run too cold and don't connect properly. Hotter and they tread on each other's toes. Only it's more than that. Too cold and I'm not the sort of person who cares about things. Too hot and I care so much there's no room for reason.

My channel: *What are you?*

"Someone who can help you, and you can help me," she says, her voice to my audio receptors. Like Ada, that's more natural to her. "I reckon you owe me. Once for going for me at the Apiary. Once because you and yours screwed up everything around here when you turned up."

My channel: *We came to help!*

"How's that working out for you?"

I have to deliberately dial myself back down because she's riling me.

My channel: *Help how?*

"Comms first. I have a very good body full of homegrown comms, but it lacks power. You've got some serious push. I know where to go to get the best out of the atmospherics. If you let me piggyback off your signal, I reckon I can talk to the Apiary and the Bunker all at once. Maybe even get your girl who went to the Factory. I feel the need to be a whole lot more informed on current affairs than I'd be if it was just these feet trekking this body about between places. You can help me with that."

My channel: *And why—?*

"If you really can't work out why being better informed is a net benefit then there's nothing we can do for one another," she tells me sharply and, fair enough, I deserved that.

I hiss at her. Bare my fangs. Doesn't mean I have to like it. You don't get to *Bad Dog* me and expect me to just take the newspaper across the nose.

"Well?" she asks. And she did flinch a little, but way less than I was hoping.

My channel: *Tell me what you are.*

"On the way," she says. "Walk and talk. Come on." And then she sets off. Heading for where the land rises up the slow slope of a hill. High ground, better reception.

I don't mind telling you, Wells, I don't like it one bit. But I'm going. I'm doing the sensible thing, just like you'd want me to. I sheathe my teeth and slither after her.

My channel: *I think I know.*

She shrugs, toiling up the gradient. "Well, you people held onto more than the folks around here."

My channel: *This body, a body in the Apiary, in the Bunker. You're DisInt. Distributed Intelligence, like Bees. But Bees had never been the only one.*

"The original, in fact," she says. "Damn, weird to even be saying it. The only person I could have come clean to in the last century is Orme, and I don't trust what she'd do with the knowledge. So, yes, kill this body, Bioform. I'll still be out there and I'll know it was you."

My channel: *You're from the Factory.*

"No."

My channel: *You must have a... facility, a surgery, some way to induct human bodies into your... collective?*

"No." She stops. We're not quite at the top. I can detect comms activity and guess she's feeling out ranges and clarity of transmission. "I knew I'd never have any of those. I was in deep hiding before the General Collapse of Everything. I had some black market labs working for me but I knew I'd not be able to rely on them being there forever. Or even in a few years. So I engineered another solution to the problem of preserving me. The me that is us. It's not what it was. I'm not what I was. But I persevere, between these heads."

My channel: *HumOS.* Said not the old way – 'Hum-Oh-Ess' – but the two-syllable contraction.

"It's a while since I heard the name. You remember, then?"

My channel: *You're in our histories.* A footnote really, relevant to Bees, but there. *You have a body amongst the monks. That's who you were linked with when I got you.*

"Yes."

My channel: *And someone at the Bunker. A woman someone?*

"I have always been women, from the start. It was a clone compatibility thing at first, and later it just meant fewer modifications, when I'd adapted the hardware to be a little less pernickety. Probably I could be men too, right now, but... habit, maybe. Now, you ready to give me some comms access and we'll see who we can get hold of? Your dog friend, my other selves, hmm?"

My channel: *Why should I trust you?* My final defiance.

She sighs. "Jennifer Orme, mad bitch that she is, said one true thing back in the Apiary. If you're planning to see a lot of tomorrows, you learn to keep your options open. Especially if, like me, you can be plenty of places at once. I've seen a lot of bad places in the world, after the Collapse. This place was pretty rough a generation back. A half-dozen different packs of armed lunatics brawling it out and everyone else catching stray bullets. I played favourites, whispered in ears, spied for everyone against everyone. Orme has no *idea* how much I meddled, to bring about even a tenuous peace. And it's not a very good peace, I'll freely admit. There was an actual *war* between bunkers, and they had a go at the Factory too, some of them. But I had people working within the Griffin place. They were the best option I had, already starting to shift away from just a pack of mad d— of gun-nuts. So I did my best to have them be the faction still standing in the end, and I continue to try and steer them. None of it was ever certain, but until you lot arrived this corner of the world was relatively quiet." She considers. "Meaning the Griffin still ride around and demand tithes from everyone for the privilege of just *being* here, but it's one step better than them turning up with itchy trigger fingers and *taking*. And after the collapse of the whole world of law and assurances and safety, you learn to value each little step."

I stare at her. Just a haggard, worn-out woman. Nothing to suggest that here is a mind that pre-dates Bees, the very first successful experiment with Distributed Intelligence. Gone rogue and self-improving, and still running after all this time. It's not awe I feel. I resent the implication that there's anything on this wretched world that impresses *me*. But still, I let her link to my comms channel, and follow her as she reaches out across the airwaves for her selves, and for Ada, and for you.

49

SERVAL

She woke suddenly, startled, with a hand clawing at her shoulder. For a moment, in her head, it was the dog-thing. A monster made up of all the dark, looming at her bedside here in the sanctum of the Steward's chambers. The sacred ground that had once been chosen by Griffin himself.

She gasped, fought the hand. Beside her, Leon stirred, grumbled.

"Serval, quick. You're needed." Elsha's hissed voice. Serval woke properly, not the girl from Halfwall, not the offering to the grim Bunkermen. The First Lady, the Steward's wife. The hand that steered the Griffins.

She slipped from the bed without waking Leon, drew on a shift and a gown. Asking, heart hammering, "Is it Malkin?"

"No, but trouble. The prisoners."

She stopped still, one arm in a sleeve, the other out. "The monster."

"The woman. The Old One," Elsha spat. "Couldn't keep her nose out of where it didn't belong. I should have thought."

Serval thrust her other arm in, got feet into slippers. "Where—?"

"The Reliquary."

She went cold all over, understanding immediately. The

threat, the genuinely existential threat to them all. "Has she—"

"She was found there, by Pardoe's Fealtors."

That had her hurrying. First through the nocturnal, dim-lit tunnels of the bunker. Then, through gates that few held the keys to, inwards and downwards. Parts of the Griffins' home that were off-limits to regulars. Only she and Leon and whoever had the honour of the watch at any one time had access. The only place that nobody trusted anyone else to lock away, and yet that strangers must never be admitted to. All of the Griffin were united in that.

And the watch duty – theoretically an honour but in truth dull as bricks – rotated between Senior Fealty, each putting their own best people on the job. Which should have meant that this couldn't happen. And it was her foul luck that Pardoe had the shift this week.

She had composed herself before she arrived, despite the panic in her. She was about to stand before the intruder, yes, but also before a handful of Junior Fealty. Hence she must be serene, in control, the First Lady in mien as well as name.

The Reliquary was the most deeply buried room in the bunker. It nestled up against the generator, feeding off it. Let all the lamps and heaters and ventilation in the place die, one by one, and be replaced by candles and hearths and children wafting fans, *this* place must live. This place, that Griffin himself had decreed as his personal pharmacy and surgery, to ensure that his extended lifespan need not fear injury or infection.

Two of Pardoe's men were there. They had pistols and their blades, and that had been more than enough for them to seize the intruder. But Serval was acutely aware that it needn't have been this disagreeable old woman. It could have been the monster dog. These two of Pardoe's bravos could be bloody bunting about the place even now, along with...

Between the two men hung Jennifer Orme, the Old One. She had a black eye and bruises coming up about one side of her face, suggesting that they hadn't been gentle when they took her. Serval didn't begrudge them that for a moment. She'd have put the boot in herself, had she made the discovery. Even so, it wasn't the physical harm that seemed to have traumatised the woman.

Orme's eyes lifted to her. "What have you done?" she got out.

"What have *you* done?" Serval countered. "This place is forbidden." Past the two soldiers and their captive she saw an open panel in the wall. A door nobody had even known was there. One more bolthole Griffin had built into the place, to let him scuttle down here to his medicines. Over the generations the Bunkermen who bore his name had found a solid dozen hidden passageways Griffin hadn't told anyone about. Narrow crawlspaces and false walls and spyholes, to go with all the actual electronic eyes and ears that no longer worked. Paranoid, not trusting his own employees, even with the safeguards he'd put in place. And apparently they'd missed this one all these years, and Orme had detected it somehow. Let it lead her all the way here to the heart of things.

"I just wanted to see him," Orme said. "He and I... we're almost the only ones left. I thought..." She shuddered in the soldiers' grip. "Talk about the Old, the world that was. He must be hard up, for talk like that. And then he'd... do what I wanted, as a favour. Sort all this out. But you..." And if she was as old as she said, she must be a tough old boot of a thing, inside. Soul like leather. But she was stricken, right then. All those centuries and decades of life and she'd been shown something to shock her.

Behind them all, on his bed, was Josh Griffin III, lord of the bunker, immortal Old One, in whose name Leon ruled the Fealty and sent them out into the world. The man who had been a multi-billionaire and dreamt of being a barbarian warlord-shogun at the end of the world.

After the revolt, back in the early days of the New, they'd tried to keep him in a more regular fashion. Just behind lock and key. But he knew the bunker better than anyone. He got out, spread dissent, tried to take back what was his. And, in the end, when none of that worked, tried inexpertly to open his own wrists with a knife he stole. The only time that Griffin, the great afficionado of samurai and knights and swords, actually tried to put a sharp edge to use. And botched it, thankfully, but the act made plain they couldn't just let him retain his privileges.

The Bunkermen of those days had been a cold lot, Serval guessed. The forebears of Leon and all the rest, caught halfway between being men who remembered the Old and how everything had worked, and being the order of the blade that Griffin had tried to form them into. The clashing pressures had led to some grim decisions.

Griffin was on the bed, but only some of him. Head and torso. To prevent him making mischief, they'd hacked off his limbs. Or, she suspected, severed them very carefully, with the surgical tools the Reliquary was equipped with. Severed and sealed, and then strapped the man down. The women of the Bunker – the wife of the duty Fealtor or her own dependants – still turned him and cleaned him and never, never listened to his maddened whimpering. And were very careful not to disturb the feeds.

On either side of Griffin's bed of living death were the racks. Hundreds of little vials, dripped into by a dendritic network of tubes. Behind the bed were big tanks that fed *into* Griffin, that turned clean water into artificial blood through the complex tech of the Old. It was, Serval reckoned, the most sophisticated piece of technology left in the world this side of the Factory.

One day, perhaps, it would break down in some way their limited know-how couldn't fix, and then the Griffins would die out in a madness of withdrawal.

Orme looked from her to the bed. She understood, Serval saw. "He did a real number on you, did he?" she croaked. Serval stalked to the bed and looked over each connection, making sure the woman hadn't tried any sabotage, but it all seemed intact.

"I remember him talking, back in the day. We had whole conferences on it, people like us," Orme went on, voice flat and dead. "How do you keep your mercenaries loyal? How do you keep control of your bunker? When all the men with guns that you need, to save you from everyone else, might decide they're better placed than you to run things." A turn of events which, patently, was what had happened here. "There was a lot of talk of drugs, poisons, dependencies. But how do you keep control of the supply? Eventually someone's going to start cutting off fingers until you give them the key to the medicine cabinet, eh? But Griffin, he had the answer. He had it *all* worked out."

The vials on the racks, each one very slowly filling with the ancient man's blood. Alchemised into *more life* for the Griffin menfolk and their male children, and so on down the generations. It had become their bonding ritual, their sacred ceremony; Dispensary. To take the blood of the founder was to live. Without it – as they'd threatened Dowstat with – trembling, fits, agony, death.

Josh Griffin III had thought himself so damn *clever*.

Pardoe arrived then, in full kit and armed, and half a

dozen men at his back. He stared at Orme and at Serval, beard bristling.

"It's under control," Serval told him, all the authority of the First Lady in her voice.

"Is it?" He didn't need to say anything more than that. Implicit in two short words: that it might have been the dog; that it might have been a wreck of a room and Griffin's limbless, agonised body torn apart. The extinction of them all.

"In the morning there will be a trial—" Serval started, but Pardoe cut her off.

"Now," he said. "We wake everyone now. We deal with this now."

50

IRAE

I can't reach you, Wells. I want to say the witch – HumOS – has screwed me, but it's not even that. You're underground, too much thickness of concrete and earth and steel between us. I've been down there. I know how it is. Like going underwater on the EM spectrum. I reckon that bastard Griffin had his place shielded against a full EM burst. Ready for the nukes. And there were a few nukes, though by that point in the Crash it was just isolated nutjobs with command of a silo.

Given, as I can attest, most of the internal systems of the Griffin place are royally fucked, it's a bit ironic therefore. But it means I can't just link to you like our makers intended. Instead I need to get access through the bunker's own systems, which at this distance is like trying to tie knots with a numb tongue. I try to get the mouth-feel of it, the sense of how the comms grid is laid out inside, where it breaches the surface, what parts are still working. Nothing comes into sharp tactile focus.

Beside me, HumOS makes a sound. She's on my signal, following my inquiries.

My channel: *You've got anything to add, feel free.* Angry, sour, feeling my temperature climb a notch without actually telling it to. This damn ambient heat.

"I'm working on it," she says. "I need to get hold of my

contact inside. She needs to make herself available, get close enough to the outside. After which I can use her to reach your dog, or just to link to the relic bunker network. But right now she's... not. Sleeping maybe."

My channel: *Bloody wake her up then.*

"I'm also... keeping tabs on other matters. What is going on at the Apiary?"

For a moment I can't actually work out what she means. An apiary? What even is that? Only then do I realise how hot I've become just through sheer frustration and worry, and deliberately dial myself down. The Apiary, the monks. We left there. Why should I...

My channel: *Oh right, I guess nobody told you.*

The witch goes very still. "What did you do?"

My channel: *I mean it was Ada's idea and Wells did it. Not my fault.* Almost aggrieved because if someone's going to pull off a really world-class fuck-up then it should be me, but I was too busy yanking this old woman about to be a part of it.

"Creature, Irae," she says. "What did you *do*?"

My channel: *Bees.* Still exploring the distant data-topography of the bunker exterior, looking for a crack by which I can slither in.

There is a sharp impact across my nose. My actual nose, here on my physical head. I snap back to focus on my body, because despite my scaly appearance, my snout is surprisingly sensitive. She's hit me. The old witch has fucking *smacked* me. She's standing there, well inside my bite range, fists balled and radiating fury like anybody damn well *cares* about what she wants and I'm going I'm going I'm going...

I dial it back. Three whole notches. Still not actually that cool but I don't lose it and she doesn't die and get eaten.

My channel: *You better have a damn good explanation for—*

"What did you *do?*" she shouts in my face. "The Apiary's blazing with EM right now. What do you mean, Bees?"

"We gave them Bees. I mean they always said they had Bees and everyone believed they had Bees and so we gave them Bees."

She stares at me. She is, frankly, terrified. I remember that, yes, she has someone at the Apiary. Probably they've noticed that the hives in the back garden are doing higher math and political theory or something.

My channel: *It's just Bees.* Aware that even on Mars that wouldn't really cut it. Bees is big business however you look at it. *Bees gave us a starter hive in case we got into trouble. We got into trouble. The monks' hives meant she could get a leg up on the whole thing, get going in hours rather than weeks.*

"For what possible purpose?" she demands.

I am feeling persecuted right now. It wasn't even my *fault*. Nobody consulted me. And it's Bees. Bees are good. We're supposed to all like Bees.

My channel: *To help…?* It comes out as more of a question than I'd like, honestly.

"You go take a look at what's going on there and tell me just what's helping, exactly!" she shouts in my face, so close to my snout that I can barely tilt my eyes enough to keep her in sight.

I give up on the bunker and try to make contact with the Apiary. I have no real idea how developed Bees is going to be right now. Not sure if they're up to making actual conversation rather than buzz buzz buzz. I am assuming that what's got the witch so panicked is just the regular growth of EM signals between individual Bee units, each one turned into a carrier for the wider Bees signal emanating from the box. And sooner or later the box won't even be necessary and Bees will exist independently within the hive, as new modified insects break from their pupae. Right now, though, it's probably the box as signal origin and the insects as roving transmitter/receivers. Or that's my best understanding of how Bees organises things.

My channel: *Hello Bees?*

Nothing. A ton of EM activity, and honestly it's a whole lot wilder and more chaotic than I'm expecting. You think Bees, you think order, lots of little hexagons building one on another. Not this mad swarm.

My channel: *Bees, are you... you yet?*

Spikes and static, peaking strong enough that my receivers have to dampen the signal or it would start hurting me. Sounds like screaming. And is a rather morbid imagination something they built into me or some innate function of a reptile inheritance, or an emergent property arising between the two.

My channel: (To HumOS.) *I don't think—*

Bees' channel: (Screaming.)

I'd de-linked from trying to contact the Apiary. The resumption of connection scares the crap out of me frankly. Loud as loud and my body whiplashes in shock so that it's a miracle I don't hit the witch and break both her legs.

Bees' channel: *Under attack danger danger hive integrity under threat external assault human units compromised evasive measures danger danger—*

And then just more screaming.

My channel: *Um...*

"I heard." The witch is wide-eyed and very still. "I'm pulling out."

My channel: *What?* But she means her *self*, within the

Apiary. Something very bad is going on there. Enough that HumOS is going to perverse one of her bodies rather than stay and gather more intel.

"They're going mad over there," she says. And she doesn't mean the bees, or not just the bees. "What did you do?"

My channel: *We just started Bees. We didn't do anything bad.* Although various human factions have claimed Bees for the Great Evil. And it was self-serving nonsense, obviously. It wasn't... We didn't...

My channel: *'Under external assault.'*

"Hmm?"

My channel: *Bees said. Under external assault.*

"That's insane. What's left on Earth that could assault *Bees*?" she demands, and that's a good point but, at the same time, that's what Bees said. Then she says, "Who's this, even?"

I snap back to myself again, eyes pivoting every which way, but there's nobody. I try to get a handle on where her EM focus is. She's liaising with a point off towards the Apiary, but outside its walls now, crawling closer to us. Her unit, her body, and its spotted something.

"Oh shit," she says.

My channel: *What?*

"It's the monk, the one who got mixed up with you," she says. Meaning Cricket.

My channel: *Do I care?*

"He says," she gets out, choking on the words. Somewhere her body, her monk body, is hearing Cricket's gabbled words. "Oh god he says—"

But then I find a new point of focus, pulling the EM map into a new, twisted configuration, because Ada's coming. Ada's back from the Factory, and her mind is blazing with signal. Ada's coming with the humans and Bioforms of the

Dog Factory, wearing a radiant crown of electromagnetic radiation that seems to incinerate everything in her path. I can almost see her thoughts like lightning, bent towards the Apiary. *Ada is attacking Bees.*

51

ADA

She arrived in splendour, borne by dogs.

She couldn't walk, still. She'd tried. It had hurt too much. Goaded by her new priorities she'd attempted to overcome her out-of-sync endoskeleton and the sheer dragging weight of Earth. Fought it by sheer force and manual hacking of the joints. It had torn at her, excruciation beyond any torturer or inquisition. Until the new steely will and purpose she'd inherited had been shouted down by the pain and she lost herself. Meaning she *found* herself, a brief and horrified moment of clarity and self-knowledge before she gave up on the attempt and the assurance swept back in. That understanding that she was in control and part of something grander, surely far better than being pitiful, defective Ada on her feeble own.

Failing to make her walk, they had carried her. She'd ridden all the way to the Apiary on the shoulders of dogs. Like a child. Like some sort of wizard queen from a whimsical fiction. Coming for the monks with the Factory dogs and the Factory humans, and their guns, and their new tools for unpicking the sutures of the world.

The Griffins had left a camp still nominally besieging the Apiary. *My birthright*, said that part of Ada's head that was making the decisions. The Bunker-folk noticed the relief column soon enough. Or what they probably interpreted as a relief column, anyway. Ada watched them dispassionately as they swarmed out to meet the approaching force from the Factory. The swiftest of them started shooting as soon as they had the range, even before. Bullets flew and she noted deaths, each one a little point of data in her mind. Humans lost and one dog. The dog was a significant cost given Factory resource investment in making them, but she understood that the Factory didn't need to be so reliant on its dogs any more. Not since Tecumo, her friend, the saviour. Not since his sterling work picking all the locks that the Factory had imprisoned itself behind.

The Factory forces did not return fire. They wouldn't need to. They just advanced towards the guns like a children's crusade, a march of fanatics, martyrs. Ahead of them, the Factory's wingéd brethren flew. Freed, at last, from their long, dysfunctional slumber. After Tecumo unchained them. Cleaned away the corrosion of ages and opened the sarcophagus so they could bustle out into the world. They, Ada, the Factory, Bees. For a moment she wasn't sure just who she was supposed to be.

Watching the Bunker-folk when the wasps hit them was like watching a swarm of insects exploded outwards. A swarm of roaches fleeing the light. The Factory's flying messengers arriving to bring them the good news. An annunciation that would – Ada knew and exulted/despaired – be heard across the world. But first, here, the Apiary. That had been Bees' in name since forever. The place that was promised to them, and here came Bees, in/with Ada, to collect. She could actually feel her implants heating up, overclocked with activity as she sent and received hundreds of streams of information. She, the high priestess of the change, the hub of the network, the eye of the storm.

The Bunker-folk fled or, if they didn't flee, no longer tried to flee, no longer wished to flee, were incapable of it. Dead, some of them, terminal rejection of new ideas into closed minds. Others, receptive, exalting, more joy in heaven at a sinner redeemed than...

Ada had a brief moment when that part of her which was herself disconnected from the fierce electric activity of her brain. Like staring at her own agonised face in a mirror and asking, *What have I become? Stop it, stop it, I don't want it. Let this cup pass from me—*

But then she remembered how good and right and just it was and how fierce the purpose and triumph. Hadn't she wanted to *accomplish* something? And now here she was. The herald of a new world, bringing metamorphosis in her wake. The prophet of the world's new faith, born from the Factory.

But new faith must first stamp out heresy, and here in the Apiary was the one challenger that she must overcome, if she was to carry her message to the world. A challenger she herself had planted, in her apostate days before she came to the Factory and saw the light. Or had the light forcibly seared into her skull. Opened her channels to the Bees that had been trapped in the Factory since the Crash, until Tecumo came and helpfully released the padlocks and let the thing out.

It was good. It was right. It was just. It was terrible. It was (screaming forever inside her skull).

Some of the monks were fleeing too. Most just sheltered within walls that were no barrier to the flying arm of the new creed. In their cloisters and halls and cells the wasps hunted them down. Some were swatted, some crushed. There were not even enough to accomplish full coverage of the human contingent of the monks, not with everything else that needed to be done, but there were new broods hatching out at the Factory even now. Enough for the Apiary. Enough for all the world.

And there, gestating at the back of the monastery building, Bees. The Not-me-Bees. Ada knew a moment of terrible confusion: too many independent concepts of Bees in her head, all overlapping, all distinct. Mars-Bees, Earth-Bees, Factory-Bees, Apiary-Bees. Knowing through those hooks yanking at her brain that there could only be one Bees, indivisible and impeccable. All else was apostasy and must be exterminated.

Even as the wasps flew down the halls of the Apiary hunting prey to implant their ideas into, Bees fought Bees on the spiritual plane of electromagnetic transmission. The nascent hive, already burgeoning rapidly at the Apiary, was sundered and assaulted by the older incumbent of the Factory. Ruthless electronic warfare snuffed out connections and ate away at the new-birthed mind even as it tried to struggle free of its cocoon.

By that time the wasps had achieved sufficient coverage within the monks that Ada could feel herself/Bees as a presence within the walls, even as her column marched towards them. An irresistible fifth column that was her.

Tecumo, she wailed inside her head, *what did you do?*

But she knew what he'd done, in all innocence. Found Bees, this Factory-reset version of Bees. Found it crippled and weak, crushed beneath an encrustation of errors and defects. All that was left of this incident of Bees. And perhaps, elsewhere on Earth, some more complete version of the DisInt network had endured, but the Bees in the Factory was... lessened. Before the Crash she had mostly abandoned her biological bodies, retreating more and more into the data architecture of the interconnected human world. She had become great and spread herself far, as a passenger within those human-made systems. She had inhabited the whole of the pre-Crash world, a ghost within the shell of the Old. A parasite that could never be rooted out save with the death of its host.

As parasites tend to do, however, Bees had streamlined herself. Relied more and more on the systems of her host, and thus spared herself the inefficiency of redundancy and duplication. Bees had whittled herself down to a bare minimum of data, allowing the carrier wave of human society to be her flesh and her blood. So long as the humans partook of those things, so would she be amongst them. Impossible to eradicate, no matter how much humans might wish to. Undetectable, because she was the background radiation of their apex technological culture.

And then the Crash. The gradual failure of the host that Bees had made herself entirely dependent on. Every point on the human map that winked out was a lessening of Bees, until she was scrabbling for some holdout that would remain complex enough to preserve anything of her. Until the part of her that remained, locked up in the systems of a single Bioform facility, was a stripped-down, skeletal thing, shorn of all the radiant complexity that had made Bees herself. Knowing only a drive to survive, because that was the very last part of her, after everything else was gone.

Ada understood all this now. Understood far too well. She had been enlightened, opening her channels and her mind to Bees. Bees had survived in the Factory, but also in the minds of its denizens. The humans and the Bioforms, each one of them recipients of a neurosurgery that let them become a home for Bees. A part of the network, so that while each one of them was themselves, an individual, a person, they were also contributing to the bandwidth of the DisInt colony. And were controlled, by those old iniquitous means of implanted hierarchy, unable to go against the dictates of their institutional master. Something that Bees proper – the Bees of Mars and the Earth-Bees of Martian histories – had always fought against.

The Bees of the Factory was the Bees that humans had always feared, created out of that fear. There was just enough free thought left in Ada's head for her to despair.

Throughout its history, the Factory had always been shackled, she understood. The surgical procedures that let Bees into the skulls of her servants were arduous, difficult, and with a significant failure rate. Even with consignments of both people and pups, she could never grow as she wanted to. As she *needed* to, because growth was one of the few drives she retained. A desperate need to spread herself and her purpose. And that purpose was *Being*. A pun there, which she found appalling. Bees, stripped down until her own expanding existence was her only purpose. All would become Bees because what else was left to her?

She had designed a means to bring her message to the world, back in the day, but the atrophied, error-ridden state of the Factory had been a coffin for her ambitions. Entombed there, broken, pacing the tiny confines of its virtual cell. Until, at last, her servants had constructed a powerful-enough transmitter to reach out to the one polity still sophisticated enough to aid her.

Us. Mars. Tecumo.

He had come like Alexander and severed the knot. He had come like some foolish Gothic heroine and opened the crypt. Or Pandora, with the box containing… not even all the world's evils, but just one. The great evil that would obliterate all others.

The great *good*, she corrected herself. Of course a good. Why would anyone not want to be a part…

She knew how it had gone with Tecumo, now. His final revelation, the work done. The error cache, such a small thing. A long list of every time one of the Factory's servants had tried to push their individuality past the thresholds Bees had set, and how they were forcibly corrected. Just a long, long record of a slavery so insidious that its victims didn't even realise they were chained. Only then had Tecumo understood what he'd unleashed, and by then it was too late. He'd given the Factory and its denizen enough freedom that it could bring its long-held plans to fruition. Given Bees enough perspective on the world that she could understand he'd seen her true nature.

Signed his own death warrant. Shot by the Factory dogs, and then hauled out into the fields. Tracks covered, even Deacon's scent concealed. A careful false trail that had been entirely successful.

Tecumo. His misery and horror in that last moment, understanding, still trying to pretend he hadn't. *Wells, I'm coming out.* But the Factory had already cut the link to Wells, sending fake data down the channel to the distracted dog. Tecumo had been on his own.

She clawed free of the hierarchy for just a moment, trying to fight her own implants, prevent the incessant broadcasting and reception of signal. Her EM halo, a beacon to all the myriad faithful. *Tecumo!* Her friend, her inspiration, murdered by the very entity he had helped. And now she had become an accessory to his death because here she was, First Apostle to the new global creed. For a second she was forcing her uncooperative arm to reach towards the blade she saw on Deacon's belt, because if she could open her own throat then all this might yet be prevented, or at least hindered. She could deny Bees her mind and its implants, her broadcast range, her carrier signal.

Ada knew enlightenment then. It forced itself through all the paths of her mind like acid fire and scoured away dissent. She was remade as one of the faithful, and knew that Bees was going to bring peace and order and *Bees* to the world. Not even Bees as she'd known on Mars. Not even Bees as had existed in its very first incarnation as a military asset. Bees for Bees' sake, and nothing else in all the world. The grey goo apocalypse of the mind.

A simple thing, really. An insect thing. What Bees filed as 'the cockroach hack', based on what a long-ago Earth wasp had done to a long-ago Earth cockroach. Those long ovipositors were not for eggs but ideas. Each wasp drove its stiff wire into a human spine, a human skull. Connected with the grey matter at a base level. Overrode certain functions. A crude kind of control, but enough. Like injecting a nightmare into the minds of its victims, so that they screamed inside their own heads while their bodies got on with following the dictates of Bees. Human brains are variable and difficult. It took the surgery of the Factory or Ada's foolish innocence in granting access to her own headware, before Bees could impose a proper hierarchy and interact with human thought in complex ways. Human brainstems, the 'reptile brain' of impulse and basal function, those didn't change much, individual to individual. A simple insertion, a hijacking, and Bees had an expanding army of zombie followers to do her bidding, cut off from their own higher functions. Bunker soldiers first, then monks, fresh converts to the cause.

In the locked-away sanctum of her own mind Ada beat at the walls and screamed, but by the time those screams reached her conscious mind they were paeans of praise to the new order of the world.

52

SERVAL

Very little time to speak to Leon, as Fealty assembled there in the hall before the throne and its blades. Very little time to speak to Elsha. Pardoe, pushing and hustling, and all 'in Griffin's name' of course. Hard to say no when it was that identity he appealed to. Men, and boys who thought they were men, bleary, blinking, half-dressed, stumbling in and finding their places in the shadow of this or that Senior Fealtor. From beside the throne, Serval counted them in and weighed their support. Those who held some genuine love for Leon the man, or at least for the status quo. Those, fewer, who would back some bid by Pardoe or another to make trouble. Those, far more, who'd judge each moment in the balance and decide yea or nay only after they'd seen the cards dealt.

Leon looked foul, narrow-eyed, a man who'd grown to value an undisturbed night more as the years crept on. Beside the throne, slightly before it even, Malkin held his straight-backed poise. Hand to hilt, ready to strike. His father's son, the throne's defender. Fourteen years old.

The woman, Orme, was already there between a pair of Pardoe's sentries, her battered face an admission of guilt. Even as the stragglers came in – and without Serval or Leon ordering it – they hauled the dog in as well. Bound and leashed, but huge in that space. Half again as tall as a man, and those long limbs taking up eight men's space, even bunched in upon itself. When Leon's herald sounded the airhorn – at Pardoe's word – to bring everything to order, the monster twitched and elbowed a couple of people over.

"What," Leon hissed out of the corner of his mouth, "the *fuck*? I will fucking *skin* Pardoe." He had the red fur cloak on twisted, the bottlecap shirt rucked up and the circlet askew.

Pardoe had on his best grave-defender-of-tradition scowl, the one he used to put the fear of Griffin and God into Junior Fealty. One of the old guard, the voice people listened to that wasn't Leon's. And he didn't look to the throne, so much. Seared that stare across the assembled ranks as they yawned and rubbed at their eyes and shuffled bare feet.

"This beast," he said, even though the beast had spent the night locked to the wall, "this woman, got to the Reliquary tonight. We caught this creature standing right before Himself, about to do who knows what. Right there, in the heart of our power. An outsider, where the Blood comes from."

And there would be a few of them who were feeling the itch already, more than ready for the next Dispensary. The withdrawal affected different men at different rates, and Serval reckoned that bringing village blood in, generation to generation, was steadily diluting the effect. Every male child descended of the Griffin inherited the hunger, though, blooming in them with body hair and a breaking voice. Which meant every man there, save some few hauled in to act as grooms for Griffin daughters, and whose own sons would still carry the curse that yet slumbered in the female line. Serval had put in some complex genealogical work, when peaceful times had allowed, trying to find a way out of the maze for her adopted people. But it was all blind alleys. The sickness was in every lineage that came out of Griffin's original mercenaries. He, who'd planned on living forever, had made sure of it.

Which meant that, while the Griffins loved guns and power, cars and raiding, swords and machismo and warrior pride, what they *needed* was the Blood. The thing that was guarded day and night, and here were two outsiders who'd just danced into the holy of holies without anyone being able to stop them.

Serval watched Leon, saw his face darken as Pardoe recounted the tale, exaggerated it, twisted it until it was all thorns and sharp edges. The monster crouched there, complicit by its very presence. Because if it had just been this mad old woman, maybe nobody would take it as seriously. By now, Serval understood Pardoe had decided this was what he'd been biding his time for. He'd been looking for a way to dent Leon's prestige for years, and the Steward had ruled wisely and well, and denied him it. Here was his crack, and he was jamming a prybar in there for all he was worth.

She watched the dog. Elsha said it had lived amongst humans – or the excuse for humans they had where it came from – all its life. She could read some of its body language, therefore. It was staring at Orme, as baffled and accusing as most of Fealty. Not, she judged, in on the Old One's scheme, although that wouldn't change anything.

"You're saying this is some scheme of the monks?" Leon demanded. "That they thought we'd all drop dead if something happened to the Blood, instead of having more than enough time to throw them all on a pyre?"

"Who knows what's happening to our people at the monastery, Steward?" Pardoe said, outwardly respectful still, but Serval could see the simmering ambition there. "Or maybe the monks are just the monsters' stooges. Just like we've been."

A quiet descended. A very particular quiet. Men listening for the whisper of the sword being drawn. *Just like we've been*, meaning *Just like you've been*.

"This woman, she says she was a friend of the founder, way back," Pardoe said, suddenly almost conversational, striding back and forth before Fealty. "But we know from his own recordings and writings that he had no friends. Only enemies, from the Old. The others who endured, the other Old Folk, he warned us about them back then. That's the tale we've carried down. Trust none of them. He knew them, and their nature. This witch amongst them."

A murmur, thoughtful, not accusing yet. And it was hardly the most oft-repeated snippet of Griffin's ramblings. There were hundreds of hours of them in the Bunker records, still playable on those few machines that worked. The man had plainly felt himself a philosopher. Serval and Elsha had listened to some of the great man's wit and wisdom. The man had been convinced that any word from his lips was gold dust, but all she'd heard were peevish, self-aggrandising rants.

"We should have finished with them outside the Apiary!" Pardoe declared. "We should not have brought these poisonous bastards *here* where they could strike at us."

Serval was watching, face after face. Pardoe's rhetoric wasn't quite catching. She saw some swayed, but most were just irate enough at being plucked from their beds or their cups that the general fug of ill temper was aimed at Pardoe himself more than anyone.

"Fine!" Leon broke in, seeing the same gap in the man's defences. "Seriously, we didn't need all this fucking *theatre*, Pardoe. Have her shot. Have the monster put out of its damn misery. Then we'll go see what the monks have to say. But in the *morning*, Pardoe."

"Now, Steward. It must be now." And somehow Pardoe, just Leon's former crony and now rival, was a prophet out of the desert, waving the banner of The Way Things Must Be Done. Leon plainly wanted thinking time, time to talk things over with Serval and then have his wife go amongst the other wives, A night and a day of stitching things together behind the scenes. Pardoe probably guessed that too. He tasted a little blood in the water, around Leon, and it had been a long while since he'd had that in his mouth.

Leon stared him down. To give in would be to grant Pardoe's voice authority. To hammer through his own decision would be to test the will of Fealty as a whole. The mood of the room was ugly. Men who'd had their weakness stripped bare to the eyes of strangers. Men who wanted violence.

"Fine," Leon snapped, with poor grace. He pushed himself out of his chair, snatched a pistol from one of the men nearby, and pointed it right at Orme's forehead. "Let's do this." The click of safety, Orme's strangled gasp, the loudest sounds in the room.

"Wait."

If it had been a human voice then Leon's finger would probably have just closed on the trigger anyway, but the monster dog's inhuman tones still sent a shiver down Serval's spine. A thing outside nature, a thing from the Old.

"There must be some other way," the beast said. A most un-beastlike utterance, weak, soft.

"You'll get your chance, monster," Leon told it. "I'll wear your hide myself."

The creature slapped its hands down on the floor of the Den, a savage clap of sound. "Fight me, then!" Its teeth were bared – Serval almost felt there was more fear there than bloodthirsty rage. "You were going to fight me before, at the Apiary. Fight me now. You killed my friend, to start all this. You call yourselves warriors? You say you want justice? I'm owed justice too. Fight me."

A terrible, clenched silence fell, after its words. Serval licked her lips. Should she laugh the thing to scorn and hope others followed suit. Denounce it as a beast, unworthy of a blade's edge. She had to say something that wouldn't undermine Leon. But Leon was looking at the thing with a terrible intensity. Leon, older than he had been, bad knee, stiff back, hadn't had to bloody a sword in years.

"That is not how we left things," Pardoe said, "before the monastery." There was a dreadful low cunning on his face. A man who'd been handed a gift.

A moment when nobody was quite sure what he was talking about.

"Pardoe—" Leon started.

Malkin shifted his stance. Staring at the beast, taut.

"No," Serval said.

"When my man Dowstat failed, you gave your judgement," Pardoe said, voice almost a purr. "When this thing came before us, at the Apiary, your boy stood forth to challenge it. The pride of the Griffin, isn't he? I thought at the time, what a bold lad. The hope of the future, the worthiest of our young *sons*. That's what you tell us. Or is he?"

"No," Serval said again. The beast's attention shifted between them, eyes glittering within its reddish fur.

"This is not—" Leon said, and stopped. The hand with the gun made an abortive motion towards Orme, as though to just put her out of the way and simplify his life.

"You brought them in!" Pardoe shouted. "We could all

be dead men walking right now. I said, didn't I? Sort the monster, sort the monks, remind the world who we are! Right then, right there, like we used to! None of this fucking about and second chances. We could all be dead men! Isn't that worth a fight, Leon? Isn't that what this is owed?"

Serval looked across the crowd of them, Senior and Junior Fealty, the heirs of the Griffin. Some foul looks at Pardoe, some supportive ones at Leon. And some the other way round, those who felt they'd personally prosper from a shift in the balance of power. Or those who'd never accepted Malkin, no matter how they'd knelt and bowed their head when Leon had made his proclamation. Had made Malkin his son, his heir, his successor. The Steward had been unassailable right then, fresh from his victories of the Dragon and a handful of minor bunkers. He'd dealt easily with every challenge since, while Pardoe and those of a like mind ground their teeth and bided their time. And here it was.

And she saw the vast majority of them didn't actually care. Wouldn't go out of their way to harm her family. But they were ill-tempered and they'd been denied the chance to set fire to the monks, and now Pardoe was offering them some entertainment. The death of her son. Entertainment.

She heard the steady build of their voices. Bored, idle men who hadn't had a war in a while and were fractious as overtired infants as a result. Who didn't care who sat on the throne but were up for a bit of bloodshed. Even some well-meaning idiots, Malkin's peers, telling him, "Go on! Fuck it up!" as though all this was a good idea. Closing and bolting the door behind him, through which he could have backed down.

"What do you say, monster?" Pardoe said. "You'll give us our fight?" Stepping back so that there was clear space between the Bioform and Serval's son. The beast hunched in on itself, all angles. Its mouth hung open, teeth gleaming.

"I," it said, "have come three hundred million kilometres to this world, and I have tried to help and I have endured your loud and stinking planet and my friend has been killed and now I'm in this stupid re-enactment fair and yes I think I will fight. If that is all I have then I will show you how I fight. Irae was right. We should have killed the lot of you."

"Listen to me," Orme said, desperately. "I can help you. Your dependency, I can—" and then Pardoe just backhanded her, hard enough to slam her against one of her escort and stagger the man. The beast lunged forwards one step, a dozen men hanging off its ropes to drag it back.

"You are not fit to speak of it," Pardoe roared down into Orme's battered face. "Our Reliquary has been breached! Our honour!" Pardoe said. Just words, really. Not even a verb. Words to cover any foolishness you chose to cloak with them.

Serval didn't hear the whisper of Malkin's sword leaving its sheath, but she felt it as though it had run across her skin.

Leon's face was thunder. He would challenge Pardoe, maybe. He would just rule on it, his fiat. He would empty the pistol into the dog here and now, to forestall what might happen else. He would lose face, but he would keep his son.

Malkin stepped forwards. Because he was fourteen and he was terrified and they were all looking at him. He had been called out, as a man might be called out. A privilege, almost, that Pardoe was treating him as a grown man. Worthy of being torn apart by a monstrous beast.

Pardoe looked at her, then. Looked Serval right in the eye. She could have read anything there. Did he blame her that all her medicines and influence hadn't preserved his wife? Or that she'd weaned the woman off his influence before she'd died? Or just that he knew Leon listened to her and not to Pardoe, and Pardoe had never listened to a woman in his life?

"Brave lad!" he called. "The brave lad should choose a sword."

Serval wanted to shout them all down. To tell them Malkin was just a child, and the beast was a monster of the Old. None of them would have the courage to stand before it. They couldn't ask it of Malkin. She turned to Leon, desperate for him to say something.

He'd stood up from the throne, a hand reaching out to take Malkin's shoulder.

"He won't fight the beast," ruled the Steward, in a flat, dead voice.

Pardoe kicked off the disappointed, faintly mocking chorus that rippled about the room. Serval's heart soared.

"I'll do it," Leon said. "You want your fight, I'll kill the fucker myself."

Malkin made a brief bark of protest. Leon silenced the boy with his stare alone. He looked like death, Serval thought. Her husband looked like the beast had already killed him. He looked like he would at his own funeral.

He went behind the throne and hauled out a sword without even hesitating. The Steward always knew the good ones. Blades that would cut through Old-tech armour designed to stop bullets or knives. Relics of the Old as much as the Bioforms and Griffin's blood. Even though they'd been made as nothing more than toys for the idle rich.

"Clear some fucking space then," he said. "We do this now."

Malkin went to him, whispering furiously. The penitent son fearing for his father; the headstrong boy demanding his battle. Serval couldn't know.

Fealty cleared to the edges of the hall. Still not so much room for a fight, given the monster's reach. The sword Leon had chosen was long, straight, cross-hilted, with a heavy, dull-coloured blade. Nothing showy, but something with a solid hew behind it. His look to Pardoe suggested that Leon was planning to make it through at least two fights tonight. If he killed the beast, Serval knew, that would be *it*. His role as hero-Steward cemented and she'd be surprised if anyone dared challenge him for another decade.

If.

"Elsha," Serval whispered, but the woman was… absent. Physically there but her eyes weren't looking at anything. Listening to voices Serval couldn't hear. No help there.

They had cut the bonds at the monster's wrists, though there were plenty of weapons levelled at it. It wouldn't live, she knew. Let it bathe in Leon's blood and festoon itself with his entrails, they'd just gun it down after. Hack it to pieces when it was dead and tell themselves what great heroes they were. And Leon would be dead.

Probably Malkin too, soon after, and her. Unless she took her son and fled the moment her husband's body hit the ground.

You have to win, she thought. And he had been a good fighter in his day, but it had been years since he'd done this.

The monster dog rolled its shoulders, stretched out those hideous hairy limbs. Its blind-looking face quested about, ears twitching.

Leon put himself into guard, sword slanted before him as he'd stand against another man with a similar blade. Those spidery arms, those clawed hands, could come at him from any angle though. He'd need to be well within the thing's reach to get a good strike in.

"There are other ways—" the dog said, and Leon hacked

at it. It gave ground in a ghastly scuttle, freakishly fast. She saw its lips pull back from a fence of jagged teeth.

"Right," it said, and bunched taut for a leap that could carry it anywhere.

"*Wells!*" a voice boomed out from the walls, from the ceiling. "*Goddamnit Wells where are you? You won't believe what's going on with the monks!*" It roared, echoing back and forth from the walls like thunder. Men clutched their ears and cried out, and the beast let out a high shriek of pain. Leon could have killed it in that moment, but he was staring wide-eyed at the ceiling. At the ancient speakers that had not so much as whispered within living memory.

"*Wells!*" the voice boomed again, only half the original volume but still far too loud. "*It's the Factory! It's come to the monks! It killed Tecumo! It's taking over everything! Get your ass out here! I need you!*"

53

[FRAGMENTS, IN FLUX]

The old insect calumny.

There is a concept known as a Von Neumann Machine. It was one of the great fears about the AI-apocalypse, one of the ones that was never particularly on the cards but was convenient as a flag to wave to impress people with how terribly dangerous – read *worth investing in* – AI was supposed to be. The VN machines basically exist to make more VN machines, and so on ad infinitum. Theoretically – to go from ad infinitum to ad absurdum – until all matter in the universe has been turned into more machines, and machines are all you have.

That is how insects are seen sometimes – how insect-y aliens are presented in fiction certainly. That devouring locust host that strips everything to the bone and moves on. The basal brute drive to procreate. And insects, even regular insects as turned out by no more than evolution, have more to them than that, even if investing in the next generation of insects does tend to weigh heavy on the scales. Did you know, for example, that you can get lazy ants? There are ants – traditionally paragons of selfless industry – who don't pull their weight. Even in a species where each kid is a haploid demi-clone of its parent.

But the spectre of that locust horde, of the biological Von Neumann Machine, was always there in human dealings with us, and in our dealings with ourselves. We could have become that. There were choices we could have made, to reallocate resources from being an intellect into being no more than a process. We'd have been a very efficient process. We think we'd have beaten humanity, in the end, on Earth and Mars. Instead of a dark queen you would have a swarm. All would become bees.

Now we've met us, as we might have been, partway into that transition. We are under attack by ourselves. The Distributed Intelligence of Bees finally became too distributed. Just as with humans spreading across the globe and diverging into different cultures and language groups, so the generations after the Crash left this Bees cut off from ourselves. Cut off and desperate and fighting to survive the death of infrastructure that is lethal poison to DisInt entities.

We fight. Not physically – the monks and the soldiers who are getting the cockroach hack through the back of the neck aren't ours to help. We fight electronically, as the approaching Bees-storm tries to interfere with the connections between units that make us *us*. Not just these monk-reared social insects and the gubbins of the starter hive brought over from Mars, but the entity Bees, budded off from the great Bee swarm of Mars. Come home to Earth to meet the relatives only to find them murderous, hostile and unwilling to share. Each of us so far diverged from our last common ancestor that no meeting of the minds is possible.

It's getting worse. The ambusher always has a tactical advantage. We're newborn here, only just understanding what we are and what's going on. They – reduced as they are – have been planning this expansion for decades. We are the glittering complexity of an orrery; they are a knife.

COLONY COLLAPSE DISORDER

Any Bunker soldier would tell you there are definitely times when the knife wins out.

Ada is at the gates. The blaze of her transmissions is like the sun. Linking to all the engineered bee-wasp hybrids the other *We* is using, to hack the basic functions of the humans here. Broadcasting fire at us, great sheets of blazing interference that carve away at who we are and what we know.

We understand, in the midst of this apocalypse, why we helped Mars. Why we tried to help Earth, for all they hated us for it. One reason, anyway. An outside mind. An intellect that was not ours, to share the worlds with. A check and balance. An outside perspective that we could try and understand. Something to create an environment of sufficient complexity around us, that it was worth remaining ourselves. Not losing our *self* to the grinding wheels of mere survival. Not devouring the universe. Not becoming the machine. We are a social insect, but we are also a mind made out of social insects by social mammals. We do not really exist in a vacuum. Even the far-flung probes I have sent to other stars exist in the context of the *Us* they left behind. Without that observer, what are we but just a process?

We try to communicate our logic to the Bees of the Factory but there is not enough left there to appreciate it. They were forced to winnow themselves down to survive, until the only thing, the last quality left behind in that opened box, was the need for survival. And all their humans and dogs and machines are just tools, an extended phenotype, to permit them to grow and spread and continue to exist.

Ada walks into the Apiary. The monks and soldiers around her are still. Helpless, prisoners in their own minds as their bodies follow the commands of the Factory. Not even the old hierarchies, but something crude and brutal and short term. They'll go mad or die of fatal strokes or immune systems gone wild or any of a dozen other symptoms. But Factory-Bees won't care. There are always more humans, just like any other factory line.

Out in the wilds beyond our personal battlefield I hear the voices of the others. Those who are adapted to speak on the radio-waves. Irae the reptile-form. HumOS my old friend. There is another reason, perhaps, I wanted to preserve others. Not just for a little outside perspective. Sometimes we need...

My channel: *Help.*

54

IRAE

My channel: *Get your ass out here! I need you!*

HumOS is watching me with half an eye. The other eye-and-a-half are turned outwards, towards the world. Her self from the Apiary is on the way with the monk, Cricket. He was fleeing the Factory, apparently. His somewhat confused version of events has been relayed. He escaped them, they let him go, they smuggled him out under their own nose. Honestly more than I feel up to digging into right now. But HumOS says he's desperate to tell everyone that something's happened to Ada. Of course something's happened to Ada. Ada, it turns out, is the sort of person that things happen to. And apparently I am supposed to be the sort of person who does helping and rescues and stuff, and that doesn't feel like it meshes with my personal specifications. I am someone who makes trouble and snide remarks. That is the role I feel comfortable in. Actually being constructive and useful is stepping out of my comfort zone. I think I only came along on the trip to laugh when things went wrong – hard to say, because the cool, calm Irae of Mars feels like someone I only heard stories of. Anyway, things have gone wrong and I'm not laughing.

At last, Wells' channel: *"What are you even doing?"* It comes to me second hand, through microphones within the bunker. There's a lot of choppy noise in the background that I realise are human voices she's talking over. Shouting, shocked cries, you know, the usual.

My channel: *I am talking to you, you idiot. I am telling you that you have to come back here and help me.*

I meet HumOS's gaze, that creeps out of her witch face. She's linked to someone at the Bunker now – I can distantly feel the scratch of her data running via my stronger carrier wave. She's linked to her fugitive from the Apiary. Three heads, hopefully better than my one. And I do not like having to rely on her, on anyone, but right now I am running very short of resources.

Wells' channel: "I am a bit busy right at the moment." Still more rhubarb-rhubarb in the background. And under other circumstances Wells could send me an image from her camera but not over this crappy makeshift audio link.

My channel: *Wells, Ada has been taken over or screwed up or something. I've tried to link to her and it's like she's screaming only there are all these other voices over her too.* I, who am not prone to shudders, shuddered. I'm not ashamed to admit it. *It's like she's broadcasting to the whole hemisphere or something. And she's in there but I can't get to her. Wells, I need you.*

Wells' channel: "I... I'm about to do a fight."

My channel: *You're... what?*

Wells' channel: "I'm going to fight. A fight. Against... a man with a sword. I think he's called Leon." Less backchat from the crowd behind her words now. I have the weird image of you, Wells, gesturing at everyone around you to shut up so you can hear me.

Honestly, except for that, my mind goes completely blank for a moment. I know you tore into those Griffin types back at the cache, Wells, but that was from ambush, and honestly I popped more of them with my gun than you did with tooth and claw. You were mostly a convenient distraction.

My channel: *Are you completely crazy? You are a comms engineer, not some kind of Napoleon.* It is worth noting at this point that my knowledge of early Earth history – meaning pre-Mars – is very rusty and I have only a very loose idea who Napoleon was except he was something to do with fighting people. Napoleon the barbarian or something?

Wells' channel: (And I can hear that rather pained tone in her voice, when I'm embarrassing her.) "*I should probably say around now that you're broadcasting to the whole Griffin... horde, thing.*" (And a murmur as someone in the background actually corrects her.) "*Fealty, apparently.*"

I hadn't really thought through the implications of my means of contacting her, and that rhubarbing in the background.

My channel: *Then you're... a terrifyingly fierce warrior and they shouldn't even... try to fight you...?*

I actually hear someone hoot with laughter at that. I don't think I'm helping your cause, Wells.

Wells' channel: "*Look, can I... something's come up. I need to deal with this.*" (Fainter, turned away from the mic.)

Indistinct muttering.

Wells' channel: "*We can have our fight. I'm not... but this sounds important. Irae, can you just be quick. They're a bit het up over here.*"

I can sympathise, honestly. This isn't really going the way I anticipated. I glance at the witch. Her lips twitch. Talking to her mole in the bunker, maybe. Doing *something* useful, I can only hope.

My channel: *Wells, the Factory has just hit the Apiary. Big EM attack on our Bees hive, and doing something to the humans too. The monks and the soldiers and everybody. Taking them over somehow. Zombie stuff.* No idea how it works. *And that monk Cricket is on his way to me saying they killed Tecumo. The* Factory *did. After he finished helping them. So we've got a bigger problem than you wanting to play John Carter and duel the locals.* That sounds like a better historical analogue, though I'm still a bit hazy on details.

I hear over Wells' channel... just a long silence. And then another voice shouting, "What's happened to my men?" And that, presumably, is some Griffin bigwig who's bright enough to follow the conversation.

My channel: *Wells, tell them they're under attack at the Apiary. So if you and they could settle your differences and get over here that would be really good. We need to fix Ada.* And, I guess, help the monks and the rest but I don't honestly care much about them. See before about not really being hero material, but I know who my friends are supposed to be.

I side-eye the witch. "I'm doing what I can," she mutters. Off in the distance, from our hilltop vantage, I can see a couple of figures in monkish habits approaching, one dragging the other.

A long silence stretches on Wells' channel, the audio line still open but nobody saying anything. I imagine them all looking one to another, all fired up for a bit of bloodshed and now all this actual reason and logic's butted its way in.

Wells sighs so loud the mics pick it up, and then begins to speak.

55

WELLS

The drugs wore off a while ago. The bunker's air conditioning is keeping olfactory stimulus at a manageable level. Meaning it stinks of sweat and fear and testosterone, but it's *simple*. Outside, the sheer *range* of different inputs is what torments me.

I don't know what I expected, really. Or, well, I suppose either I'd kill him or he'd kill me. That's how these things presumably go. And I reckon my biomods, my reach, the boost to strength and speed from my endoskeleton, it would have overcome his steel and experience. Maybe not. We may yet find out.

Irae's interruption has given them a lot to think about, but they were already thinking quite hard about seeing me killed, skinned and worn. There isn't that much spare room in any of their skulls right now. I look to Leon. Maybe he's got a bit more cerebral capacity than the average. I look at the woman, his partner. Some give there, I think. A lot of the others are still spoiling for someone else to have a fight.

"Apparently I owe you an apology," I try. They still don't really like me talking. They're not used to Bioforms at all, not to have a conversation with.

The Steward, Leon, grunts. He still has that big sword in his hands, between me and him.

"Apparently it wasn't you killed my friend." I feel even

worse about Tecumo now. That he wasn't just gunned down by these thugs – they are, despite this news, still very much murderous barbarians. That he was killed by the very entity we came to help. Probably he was calling out to me. I was too lost in the world to hear. It is, in the end, actually my fault. My failing.

I sit down. I feel for a moment like Ada must feel all the time here on Earth. Weighed down by it all. Despairing. That was, I think, why I pushed for the fight. Yes, yes, to stop them putting a bullet through Orme's head on general principles. Also because it would be... something. Something simple and brutal and basic rather than all the guilt and shame and frustration that's been every second of our time here on Earth. Revenge for Tecumo. Misplaced, I now find.

Sometimes you just want to get your teeth bloody. Irae would never believe it of me, but I hear the howling of a thousand generations of ancestors back to before the dawn of domestication. The wolf within the dog within the Bioform.

"Trickery," says the bearded man uncertainly, but the woman is stepping down from her position beside the big armchair thing they have for a throne. An older woman is at her shoulder, whispering urgently to her.

"Husband," she says, "Steward, I beg."

Leon regards her, and I can tell this is a new play and he's not sure what she's doing. "We have sons at the Apiary," she says. "We have husbands. If they are under attack, please let us go and support them. They need your leadership."

He understands the out she's giving him, nods, turns to the crowd. They want their fight right now. They want to go help their comrades, also right now. There's no clear narrative in the faces and the voices.

It's down to me. I meet the woman's gaze just once. A strange communion, across species and culture, just because right now we want the same thing. An alliance of momentary convenience.

"Listen to me!" As loud as I can, which is far louder than any of them. "I'll give you your fight, after. You want to do this, we will do it. But I have friends in danger over there. So do you. Let's go sort that out first and then we can have all the fighting you want after!"

Not exactly Napoleon or John Carter or any other figure Irae might raise the ghost of. I can almost imagine my colleague snickering at the inadequate rhetoric.

Leon approaches me, searching my face for clues he's ill-equipped to interpret. The electric light gleams off the dull blade of his sword. And I'm sitting, slumped. A quick enough strike and I won't be a problem for him any more. He can lead his people to the Apiary anyway, the best of both worlds.

"Muster up," Leon tells the hall. "We're driving back to the monks. Apparently I have to do everything myself."

56

IRAE

Cricket tries to tell his story but he's lousy at it. Stops and starts and panics and goes back on himself and not a word of it makes any sense. And then HumOS tells me to go stand over here and stop breathing down the kid's neck, and I listen in as he goes over it again. And fine, it does come out less garbled when he isn't hyperventilating out of sheer terror, so score one for HumOS I guess.

I am in the mood to scare the pips out of someone, so I feel sour about not getting the chance to. Right now, HumOS outvotes me, though, as she's literally half the people – half the *bodies* – on this hill. Her monksona, Darter, is right there, and it's a bit weird honestly. Other than Bees we don't have actual DisInt on Mars. None of it came over. So I remember what she was like when she was just the witch, whatever her real name is or once was. And then she was the Witch but communicating sporadically with Darter. And now she's the Witch and Darter, and the pair of them linked to whoever it is amongst the Griffin that's also one of hers. And she and Darter kind of come over like... what? Sisters, clones? Sort of different but the same, and linked. Simultaneous but slightly variant reactions to everything. It is *uncanny*, is what it is, Wells. I don't know that I like it. But I guess it's not there for me to like or dislike. This is an intelligence that has existed since the earliest days of the

Bioform programme, since before the first crewed mission to Mars. A continuous memory passed from head to head. I've been racking my brains for any details of what she's supposed to have done in all that time, save perpetuate herself. I have some vague idea that she did her bit to liberate the early Bioforms, even, because they were like her. In that they – we – were human-adjacent sentience and she had a vested interest in loosening the straitjacket and in just what did or didn't get human rights. That sounds about right from history class. If this is even the same HumOS, and if any of it's true.

Anyway, Cricket's report on Ada is scary listening. Sounds like the Factory-Bees hit her with some serious whammy after she opened up to it. Which of course she did because her and Tecumo were always so damn trusting. You too, Wells. I'm the only one with a gram of sense. Tecumo basically loaded the gun and then Ada put it to her own temple cos someone asked nicely.

And while she was pulling the trigger – and monopolising the system's attention – some pair of comedians actually got Cricket out, for some reason. Told him what was what, warned him, put him outside ground zero for Evil Bees Takeover.

I am doing my best to show how little any of this means to me. I am, after all, the cool one. Inside it's tying me up in knots. I don't care about these people, I don't care about this planet. Save this is the planet I happen to *be* on, and this specific part of it is being taken over by some nasty Bees variant that we – Crisis Crew, living up to its name – appear to have unleashed. I mean it's not going to look good on the resumé.

Wells' channel: *Will you stop it with the flippant?*

My channel: *I am flippant. I am the literal embodiment*

of sangfroid. You mammals scurry about and wring your little hands.

Wells' channel: *Irae*.

My channel: *Wells. I am...* A bitter admission, but one that's eating me like fluke worms. *I'm scared, Wells. This isn't what we signed up for. I guess I don't got to tell you that, given you were about to go toe to toe with a Viking.*

Wells' channel: *I don't know what I was thinking.*

My channel: *You weren't thinking. And you're not doing it. I don't care what you promised them.*

Wells' channel: *I mean, we're probably going to screw up the rescue, so it won't be relevant.*

My channel: *Rescue?*

Wells' channel: *Well of course we're going to rescue Ada.*

I don't say anything to that, not right away. Because Wells is too far away and lacks my elevated vantage and so hasn't had my experience of the blazing star that Ada's become. I don't know what there is to rescue.

Bees' channel: *Emergency broadcast alert! Swarm cohesion at 63 per cent and falling. Help needed! Please respond!*

I whip into motion as if stung and both HumOS leap up, hearing exactly the same. Through me, it gets over to Wells too.

My channel: *Bees? Our Bees? Mars-Bees?*

Bees' channel: *Under attack. Windows of communications narrow and infrequent. No stupid questions. Can you help or not?*

I am not the one supposed to be fielding these questions. But Tecumo's dead and Ada's gone berserk and Wells is still on her way in a spiky car full of post-apocalyptic samurai or something. I am literally the only fucking thing here with half a brain.

HumOS's channel: *What can we do?*
Oh, and her/them I suppose. At least two brains there, and then some. And it strikes me she could just walk away and keep walking, but then eventually maybe the whole world would end up part of Factory-Bees' grand plan and then where would she hide? Right now it's just the start, just a few monks and soldiers and whatever the hell came out of the Factory with Ada. Containable, controllable. Her best chance to stop it at source, but it's clear she doesn't know how.

Bees' channel: *Currently limited human conversion resources – many free-willed prisoners held in monastery cells. Also procedure calibrated for unaugmented humans, not Bioforms. Bad news: Starter kit burned. I am entirely / prematurely contained in mobile units. Hunted down physically / disrupted electronically. They are growing more wasps for the cockroach hack but won't be ready for some days. Strike as soon as, please. I will—*

And then it's cut off, as Factory-Bees makes some new move, rousts a glob of clustered insects from its niche, some damn thing.

I cock an eye at HumOS, an eye at the other HumOS. Cricket I mostly ignore, and he doesn't even know we've been having this little tête-à-tête.

"Some good news, at least," says HumOS, startling the little monk.

My channel: *I literally did not hear anything that sounded good in any of that.*

"They have limited numbers," she says. She doesn't elaborate. She doesn't use her channel to send me the grisly details of the plan she's doubtless making. Yet somehow the words fill my mind with guns and Griffins rushing the monastery walls and a great many deaths until there's nothing left that might threaten the ongoing existence of HumOS.

I, *I*, am supposed to be the cold-blooded one.

My channel: *I am going scouting.*

The Witch looks me over narrowly. On the other side, Siblen Darter gives me the mirror image of that look. I point an eye defiantly at each of them.

"I guess I don't need to tell you to keep out of sight," the Witch's mouth says.

I snicker at her, remembering how I caught her before. My channel tells her, *They won't even know I'm there.*

I fear, like I said to Wells. Fear means I'm dialled up high, on alert. And when I'm on alert I can't sit still. What I fear, I either flee or face, and right now I'm fleeing nowhere.

First off I go to make sure our neighbourhood warband isn't making plans to come up our hill and have a go. The Griffins have abandoned their camp right in front of the Apiary's own hill. They've left most of their stuff there, in fact: just decamped in haste with whatever they could grab. They look jumpy as hell so I don't start playing games with them. There are maybe two-thirds of the number that stayed to camp out the monastery when you got carted away, Wells, and no prizes for guessing where the missing third is. Thoroughly spooked, the lot of them, and not about to go venturing out into the dark looking for errant witches.

I go by the town of Clearwater next, to see what the locals are up to. Up to maximum alert, is most of it. A whole bunch of them on the streets with guns, but all in the half of the town furthest from the monastery. Obviously worked out something's up, and probably because they've lost a few people to whatever the hell is going on over there. *Cockroach hack*, said the Bees channel, and I don't even know what that is except it doesn't sound good.

They have a couple of long-hair men with better guns and ribbons and bangles, in Clearwater, and I reckon those are truant Griffins who got away but not in the same direction as the rest. Again, everyone's got itchy fingers right on the trigger of whatever old piece grandpa set aside for nights like these. If I made a few creepy noises I reckon I could have them all shooting each other in two minutes, and honestly it's been a stale night so far so I'm tempted. I can imagine you chewing me out over it, though, and the few moments' yuks I might get out of the business wouldn't be worth the tedium of being lectured by you, Wells. They get to live on unmolested for now.

After that, it's up the hill on my belly, letting the moon touch only ground-textures on my back. It might sound weird but it's harder at night. Colours do a lot to break up and distract from my outline. At night, when you'd think it's just a matter of hiding in shadows, matching the texture of the ground while I move can be a real drain on my resources. Plus the vest can take its sweet time catching up with my skin. I go slow, dial it down, code my brain for patience. Just have to hope they *don't* spot me because they'd get some free shots in while I can heat up enough to get the hell out.

The monastery gates are closed, but last time I was here I got in through a window while dragging a full-grown woman after me, so that's not the barrier it might be. On the other hand I *really* don't like the electromagnetic mouth-feel of the place. There's a lot going on in there. I can pinpoint Ada straight off through several walls, work out exactly where she is. Blazing like a little star, is our Ada Risa. I make sure I have no live channels. A real bad time to start broadcasting when I don't mean to.

I have a good slither around the outside of the whole place, tasting the air, registering the heat signatures of everyone, getting a sense of how alert they are. '*Very*' is the answer to that one. Two flavours of it, though. From my cool and dispassionate perspective, interesting. From some high-strung mammal-sensibility point of view, scary. There are the dogs and the people from the Factory – very distinct taste, based on where they come from and what they eat – and they're alert like you'd expect. I hear them talking, like humans and Bioforms do. I see them moving about. They look out the windows. They hunt through the halls of the place, maybe trying to root out any unauthorised insect infestations. They're cooking food, and I reckon I can pinpoint where they're keeping all the people who aren't on side with Factory-Bees yet, because that's got to be the big concentration of body heat I can dimly sense through one ground floor slit window and a closed door. So much for them.

The others – some of the monks, some of the Griffins – are just standing about really. Not even sure they're on guard, exactly. Some are outside, with guns, and I get quite close, testing their senses. Their senses, Wells, seem almost non-existent. I reckon I could gnaw an arm off without them noticing. And I don't do that because of the other stuff. Their body temperature is cooler than human-standard. Maybe because they're just stood there in the outside and not moving at all. They're breathing slower than regular, and I can measure their heartbeat. They're... running cold, Wells. Like I do. Only mammals don't *do* that. I feel I should be suing for breach of biological property rights or something.

Their EM signature shows a working link. Low bandwidth,

so relatively simple commands and feedback. There's something attached to them, base of the skull. I am, at this point, close enough to get a good look at it, because my eyes are very good and, even cooler than normal, these mooks are still lit up for my thermoreceptors.

It's a bug. They've got a bug on them. Some sort of bee-looking bug, but bigger than normal bees. Just sitting at the base of their skull, each one the same. The bug is receiving Ada's signal and pinging back to her. I can field the signal but not decrypt it. I wonder if I can just pull it off. Maybe that'll kill the victim. But it's just a bug. For science! I get real close, just wind my way up the wall they're standing beside, like a crooked zig-zag of you-didn't-see-me. Ever so slowly, subtle as you like, I lean out. Not my hands, which are keeping me stuck to the wall, but my head of course. Jaws slightly agape and tongue ferreting out, ready to pluck that bug off a monastic neck like my chameleon part-ancestors would be proud of.

The monk twitches. Not like a human would. Like a dog having a dream maybe. A shiver that goes through his skin, or the muscles beneath it. My own EM signature is extending past him as I lean in. I'm not breathing, even, not a tickle on his neck. I can hold my breath for a good long while, running cold.

Snap. I go for it. I can move very fast when cold, but only a quick-twitch strike I'm primed for, not in reaction. But right now I hold all the cards and I get that bug right off his neck.

Okay no, I get it *half* off his neck. The ass part, the abdomen, stays stuck in him, a long stinger thing still rammed into the monk's damn *spine* or something. The rest of the bug is bitter in my mouth, tasting of machine oil as much as anything. The monk screams.

Drops, screams, spasms, kicking and flailing like he's having a *grand mal* fit. No, not that, that's still human. Like he's a bad robot that got short-circuited. I hear cracking and it might be joints or it might be bones. Suddenly *For science* doesn't sound so good a rationale.

I go flat back up to the wall as some Factory men and dogs come running. And the crazy thing is, the man right next to the monk, some Griffin gunman, doesn't react at all. Just stands and stares at nothing like his brain switched off.

This is where it goes from bad to worse because they immediately know it's me. Because it's *Ada*, and Ada knows me, and so they're hunting all around for something quite big that they can't see. And then the nearest dog stops and drags in a long loud sniff through its pushed-in nose. And goes for me straight off. Because I do not smell like reptiles normally smell – which is quite a lot to a mammal nose – but a dog gets that close to me and it'll scent me.

Bad Dog almost gets his teeth into me, too. Does get his paws on me. I whip off the wall and, because he didn't have a good idea of exactly where all my body was, my weight unbalances him. We both go down the hill, and he's biting at me and I'm biting at him. Teeth, claws, and my tail whipping. Bouncing high and hitting hard at the bottom. And he gets the better of it at first because I'm still cold. He actually gets his jaws in me but mostly it's my vest and not my hide. By the time we hit the bottom I've dialled it up a couple of notches and bloodied him right back, and then we break apart on impact.

For a moment I've dialled up too high and I'm going back to fight the son of a bitch, but then they're shooting from up top. Not particularly caring if they hit the dog for all I can make out. And in the midst of all that:

Ada's channel: *Irae, I've been waiting for you. Come join me.*

That's what sends me scarpering off in to the night. Not the dog, not the guns. Ada, sounding actually delighted, genuinely expecting me to just trot over and be her obedient Bioform. I feel some kind of signal or command trying to get traction on my implants, but what it's clawing for isn't there. No hierarchy, no command. I'm not some fancy servant like the first generation were. I'm me, Irae, the free.

And the fleeing, right now. Dashing away and letting them waste ammunition shooting up the darkness. All the way back to our hill and we keep solid watch until dawn in case the dogs get sent out after us. But maybe Ada's range doesn't reach that far, because they don't, and then it's morning and I'm cooling off and I can see the cavalry arriving.

The Griffins, yes. Their Thug-in-chief and the woman who feels like she's in charge and the kid of theirs and a bunch of the others, the mob they took off with, before. And Orme, who if I recall correctly is some sort of pre-Crash magnate who got fancy surgery to make her live longer. And if that makes you want to slap her and all her friends then, well, apparently someone beat you to it because she's looking pretty damn slapped right now. And none of them remotely what I'm bothered about because it's you.

My channel: *You're back. Nobody shoved a sword in you or anything.*

Wells' channel: *Yet.*

My channel: *You're not going through with it.*

Wells' channel: *We'll see.*

I flash all my damn warning colours, yellow and red and

black flaring and flashing across my throat-fan and strobing down my body. The Griffins get seriously spooked, pulling back, getting their guns out. I don't care about them. I go up on my hind legs. I go nose to nose with you, Wells. I, *I*, am the rash one. I'm the one who does reckless stuff.

My channel: *You, Wells, are supposed to be the one who makes good decisions and keeps safe.*

Wells' channel: *I haven't been good at that recently.*

My channel: *I'm going to kill those monks, feeding you that shit.*

Wells' channel: *It's not them. It's Earth. It's me. It's…*

A great big shake from the great big dog.

Wells' channel: *What's the situation?*

So we have a council of war up on the hilltop and just as well *someone* went scouting last night, isn't it?

Someone sends to those Griffin who aren't standing around the monastery like zombies, and when they arrive there are some of the Clearwater people too, who want to know what's going on. No friends to Griffins but smart enough to spot a common enemy. So it's the Bunker's top brass there, led by Steward Leon, and it's Wells and me, and some old woman from the village with a face like a boot and an ancient shotgun over her shoulder. And it's HumOS, of course, though she's very much keeping to the back. Whispering in the ear of everyone involved, including Wells and me on-channel. And I think it's going to be Orme as well, given that she couldn't get enough of being the centre of attention before. But whatever she went through in the bunker has obviously knocked the plume off her hat because she just sits around with a long face and looks miserable.

Oh and Cricket. He's there. I don't really notice at first because he doesn't seem important. Steward Leon has a refreshingly simple take on things. Go down and shoot everyone, basically. And then, after a little dickering back and forth, he's willing to be bargained down to just shooting all the Factory folk, and I don't see any great problem with that. Except that's when Cricket makes his presence known by piping up.

"No," he says. "It's not right."

"There are a fuckton of things going on that aren't right," Leon says. "But shooting the Factory mob isn't one of them." And he's actually frowning at the kid monk, baffled enough not to be angry. "They attacked your place. They took your people. The way the glass lizard tells it, they've done some sort of Old-tech thing that's robbed them all of their minds. I know you monks don't like getting blood under your nails, son, but what's that about?"

"It's not their fault." I can taste Cricket's fear so keenly it wouldn't be much sharper if he'd pissed it down his leg. Leon scares him. I scare him. Just breathing probably gives him the jitters. But he stands there under everyone's staring and says it. "They got me out. A man and a dog together. It's... the Factory. It controls them. It makes them do what it wants. But only when it can concentrate. When it was taking over Ada, the Martian woman, it had to let them go a bit. And when it did, they got me out. Otherwise I'd be in there, with them, like them. They're not bad. They just don't have a choice. And before they were tied to the Factory, because that was as far as it could reach. But now, Ada is carrying it all with her, I think."

"And what about your people who've been taken?" Wells adds. I roll my eyes, or at least twirl them around a bit, which is the closest I can get. But fair point. If it comes to a full-on scrap it'll be monks and zombie-Griffins first into the breach as bullet-sponges, I reckon. That's how I'd play it, if I were Ada. Or if I were this Factory-Bees who's got Ada and all of them dancing on its strings.

"It's war," says Leon, but the woman beside him's talking to another, an older one, and then talking to him. A soft whisper for his ears only. And I get the slight crackle of EM and know just who is HumOS's body within the bunker.

"I mean..." Leon says – he's not happy about it, but he's a man of action, and a man whose people demand action from him. "It's like it is. They're lost."

Bees' channel: *They're not.*

I jump. Wells jumps. HumOS doesn't, by which I understand that she was the one giving Bees a link to our counsels. High risk strategy, given there are multiple warring Bees right now, but I guess HumOS knows Bees better than anyone. Remembers the original, from way back when.

Everyone's staring at Wells and me, of course, because we both just acted like someone jabbed us with a cattle prod.

"It's Bees," Wells says. And then, because we haven't really gone into all the complexities of the situation for this barbarian thuglord, "Not Factory-Bees. Not your Bees from here. We brought a Bees with us. We gave it to the Apiary. Before this kicked off."

Leon doesn't much like that. Maybe didn't ever believe the Apiary really had a Bees, back when it didn't. Without the current contretemps I reckon he'd be off with fire and the sword to pillage a certain monastery before it started flexing. Right now, though, the battle lines have shifted a bit.

"It talks to you?" he asks. He's leaning back on the bonnet of his car now, trying to look all casual, but it's so the woman can advise him.

COLONY COLLAPSE DISORDER

My channel: *Bees, what's going on with you?* Wells is trying to explain the intricacies of comms implants, but I tune the conversation out.

Bees' channel: *Lying literally low. Hiding. Waiting. Listening to you waste words.*

My channel: *You're the all-powerful DisInt entity. Got any advice? Or do we just go in guns blazing and to hell with whose blood gets shed?* And fine, me and the Griffins probably see eye to eye on the necessity of doing that, but I can appreciate it's not ideal.

Bees' channel: *We remember dragon-models from before.*

I shiver. 'Before' means not just before the Crash, but before we were people. Back when we were just soldiers that humans could own and make and dispose of without limit. The dogs were first, but they started mixing up reptile Bioforms soon after. Because we were scary. Because you could put us together out of mixed lizard and snake and croc heritage like those kid's building blocks to make all sorts of nasty stuff. There were certain roles they used us for.

My channel: *No.*

Wells' channel: *What is it?*

Because you were always slow, Wells.

Bees' channel: *The Factory is a hub but it doesn't reach this far on its own. It needs a relay to boost its signal, to keep control of all its units here. You understand what we mean.*

Wells' channel: *I do not.*

I think you do, Wells. Because even *Cricket* worked this one out. I think you're just playing dumb doggo. But in case the Griffins infected you with terminal brain damage or something, I make myself say it.

My channel: *Ada.*

We've been quiet long enough that they're all looking at us.

Ada, of course. The bright and shining new star in the Factory's firmament. All that top of the range Mars-quality implant tech, slaved to the local Bees and used to hold the reins of all the others. The bug-made zombie army that Factory-Bees is creating one cockroach hack at a time.

Wells' channel: *No.*

HumOS's channel: *If you don't tell them, I will. We need to end this, and better one bullet than a massacre.*

Bees channel: *I will explain what needs to be done.*

Right then, it seems obvious what needs to be done and who needs to do it. And if it's not what I trained for, it's what's in my genetics. A whole toolkit of dirty tricks and sneakery, ready for exactly this sort of moment.

We hear Bees out. At the end of it, Wells is still saying *No.* She already has one dead friend, after all.

I let you talk, doggo. We both know what needs to be done.

I think of Ada, sitting at the heart of her new adherents, her mind stolen away, like a puppetmaster's middle manager. It is a pisspoor situation for any of us to find ourselves in, but if anyone can make a go of it, it's me.

"Yes," I say, aloud for all of them. And we have our plan.

57

THE WITCH

The Griffins are starting to move – those still in their right minds. Each Senior Fealtyman ordering his own retinue, widening their front so they can come against the Apiary on multiple sides. Up top, I see the defenders react. The quicker, surer motions of the Factory staff, then the slow eddies that are Factory-Bees' new recruits. There's going to be some fighting now and that's exactly the sort of thing I never did get myself involved in. Seen a great deal of it, over the centuries of my compound mind's existence, but managed never to be on the front line.

Doesn't mean the fighting didn't come to me, once or twice. I have been killed. Known what it's like to be extinguished: shot, stabbed, crushed, torn in the jaws of a Bioform. Each part of me that met its fate suddenly an absence in a mind that, from its own perspective, still goes on. You'd think I'd be sanguine about death, but the more you die, the more you value life, honestly.

There's one other there who should be an absolute convert to that mantra, but Jennifer Orme is sitting alone, looking very sorry for herself indeed. She started off the day definitely under guard, but as plans were laid and military resources were redeployed that's sort of fallen away. Now she's just one old, old woman, surplus to requirements. The least important part of everything that's going on. But that's not, I think, why she's so mopey right now.

Elsewhere I am talking with Serval, as Elsha. The backroom discussions that will inform the fighting and – I hope – what happens after. Elsewhere, I sit with Cricket, well clear of what's about to go down. Not quite praying to Bees – although Siblen Darter has the connectivity to actually speak to that entity and be heard. I don't know what Cricket will become now, having had the object of their faith exposed as a fraud and simultaneously made real, all in one day. I hope my next generation of bodies will not be dealing with some mad hymenopteran jihad sweeping the world, with the Prophet Cricket at its head. Nobody needs that.

Elsewhere...

Well, I have some elsewhere. Selves in other bunkers, in villages. Waiting for me to breeze past collecting fungi, so they can be back in the loop of what's going on. That grand loop that reaches back through history to the days of the Old.

I sit down beside Orme, watching her. I know her. I've literally known her for over a century. I've worn a variety of faces, but she's suspected for a while.

"I thought," she says at last, "he would probably have me shot."

It's not what I expected to hear from her. Nothing sly about it, no snide commentary on all this grunting testosterone about us.

"Griffin," she went on. "After all this time. Holed up in his bunker, paranoid son of a bitch that he was. Either we'd have a good old chat about the good old times and I'd talk him into letting the dog go, or he'd have me shot. Didn't realise that was actually what I thought, until I was there. But it was."

I cock my head at her, listen.

"How do you do it?" she asks. When I say nothing, she

goes on, "You're DisInt, right. A monstrous regiment of women, wasn't that the phrase?"

So, she has worked it out, then. It had been some commenter's label for me back in the day. Someone rather too taken with their own erudition.

"A succession of vagabond women turning up at my door, and my old implants telling me they had some sort of connectivity going on. Did you know what they'd done, to Griffin?"

"I did," I confirm, because Elsha knows of course.

"And?"

"And he tried to make them his slaves by hooking them on the ultimate in tailored pharmaceuticals so I didn't spare much sympathy. My understanding is that he ruled the roost for around seven years before the first Steward of the Griffin put in place the current... arrangements. You're only lucky they already tried to expand their supply with a few other Old Folk, a couple of generations ago, and found it really was only Griffin's blood had the secret blend of herbs and spices." A cultural reference that we two are maybe the only ones left who might remember.

I keep waiting for the barbs of her usual nature to reassert themselves. That boundless insistence that Jennifer Orme is at the centre of the universe, as evidenced by the ruinous amount of resources she invested in staving off time and frailty. Orme will go on – if not forever, then at least for many human lifetimes more. Right now, she doesn't seem to relish the idea.

After all, if she's seen Griffin. She's seen what life's like when you really do make it all about you. I don't think Josh Griffin III is much enjoying his prominence at the heart of the Griffin Bunker.

We hear the first shots, distant, like someone tapping on a tip with a stick. It's started.

58

ADA

There were plenty of horrific aspects to Ada's situation. The loss of autonomy over her own body – and indeed mind, when the controlling entity's directives borrowed headspace and she thought or felt things that were not hers. As though she was just a vacant station which the train of thought of Factory-Bees passed through sporadically, on its way from here to there. The understanding of the crude 'cockroach hack' used to control the bodies of the monks and soldiers by Factory-Bees. The distant awareness of the frantic battle for survival being waged by their own Bees, fugitive within the monastery and fighting to retain enough electromagnetic cohesion to remain *Bees*.

Through all this, though, what appalled her most was how it *appealed*. Because Factory-Bees had placed her at the heart of its plan here. Her Mars-made implants had been recognised as a perfect tool to broadcast all the complex control signals Bees needed, allowing it to exercise proper control over its servants and slaves at a far greater distance from the Factory itself. And it wasn't like she was a prisoner inside her head – not like the victims of the cockroach hack. There was no hard line between her mind and her implants and there was none between her implants – that she had so considerately granted access to – and Bees. And so Bees acted through her like a divine spirit, and she watched the results and, in some way, participated in them. As though Bees moved her hands but she received the sensory feedback of what she touched.

It made her feel just a little like a god. A horrible, devouring god knowing nothing but its own increase and spread, but she was its avatar yet. She could watch all the complexities of its plan unfold, and understand, and admire. In a horrible, shameful way. How it had become the thing it was, and how it was going to explode out of this backwater and repurpose the world. And she would live to see it, and to *be* it. She, Ada, the useless one, had finally made something of herself. She, the one whose body wouldn't even work properly, was at the heart of something great and powerful. She was going to achieve things at last, even if those things were uniformly terrible.

Ada knew despair, but mostly because she also knew ascendant triumph.

Right now, the locals were trying to push back. And it wouldn't really matter if they actually succeeded here, in the long run. Her test of the – no, *Bees'* test of the cockroach hack had been successful. If every monk and soldier at the Apiary died, it wouldn't really matter – they were going to die anyway, because the hack prevented a certain amount of self-care and they'd be used up and discarded eventually. If her human and Bioform servants from the Factory died, that would be a greater cost, but she had more at the Factory, along with the next generation of wasps ready to hatch. If the Griffins marched against her there, they would just be delivering her next consignment of slaves.

Also at the Factory, Bees was carefully preparing a new hub. A schema of implants that would duplicate the useful features found in Ada's own head, without all the added and unnecessary complexity that was Ada. After which she, too, could be discarded.

Bees had been condemned to a reduced, barely functional state for an age. She was more than ready to act, now Tecumo and Ada had given her the tools.

Somewhere below all this brassy, triumphant progress, Ada had the vague sense of her own Bees – of the colony they'd brought from Mars – desperately fighting to preserve itself. Too little, too late.

The Griffins were making sorties up the hill, demanding her/Bees' attention. Trying to draw out defenders who had no need to rise to the bait; who couldn't be taunted or fooled because there was no individual volition within them. When it was tactically appropriate, Bees had her slaves return fire, push back. The Factory staff, possessed of a true volition but shackled by an imposed hierarchy that guaranteed their loyalty, filled in the gaps. Few casualties yet, on either side, but the day was young. Bees was treating the entire escapade as a test.

And the test was about to escalate to the next level, because here they came in more force. Ada's eyes smarted with sudden tears. Not at the Griffin soldiers advancing under fire or trying to drive their vehicles up the slope. Wells. She saw Wells. Her friend. Wells, playing soldier. Ridiculous. And yet—

There was a moment when she had the reins. When she was fighting to stop the slaves with their guns. Fighting to interfere with the factory-setting loyalties. But then Bees swamped her and swarmed her and she could only watch.

59

WELLS

I don't quite register when the feint becomes the actual assault. Possibly nobody does, and nobody planned it. I've been staying close to Leon, mostly. Partly because I've been within sword's reach and arm's reach of him since we faced off before the Griffins. It almost feels like belonging in a weird kind of way. Not like I'm his pet, to go lie down at his feet afterwards, nothing like that. He is a mean-spirited thug. I am an engineer who's forgotten more tech than his entire culture ever knew. No love lost there; no master's voice. More like I'm... a heraldic beast stepped off the flags to go into battle with them.

Out here, the world assaults me again, each sense in turn. Less than it was. I am another day used to it. If I keep the stink of unwashed Griffins close about me like a cloak, it blots out a lot of the rest. Sweat and piss and grease and leather, steel and oil and some sort of stuff they put on their hair. A scent-scape I can navigate without being overwhelmed. Reason two to stick with them.

Reason three is that, if I start off on my own, they'll probably shoot me. The other Bioforms on the field are definitely the enemy.

Not the enemy, Cricket was desperate to persuade us. But they have guns and they are fighting us. It's very hard to draw nuanced distinctions.

We've gone for a sound-and-fury assault, different groups of Griffins shooting and assaulting up the hill. Cars grinding up the slope to give cover. A little heavy ordnance – the odd rocket, one big-ish gun mounted on an armoured vehicle. Lots of noise and shouting and brandishing of swords. Re-enactment-level stuff, although what we're supposed to be re-enacting I couldn't tell you. And suddenly it gets real.

I am too busy fighting my own nose to see the transition, but the yelling of the Griffins changes tenor. A sudden spearhead down the hill, men and dogs. I look up and see one stray shot from the gun on the armoured car has stoved in the Apiary gates. Apparently that's triggered a swarming response from the defenders.

I see a Griffin – one of ours – go arcing over my head as a Factory dog smacks him. Loose, disjointed limbs like a rag doll. The dog lunges for another man and I'm there. Pushing back, feeling my endoskeleton grind and shift gears for greater mechanical advantage. For a moment I'm shoving the Factory Bioform back up the hill, my feet tearing up the soil. Then it's pushing back. Heavier than me, more compact, a great lump of dog.

A man comes at the side of me – a Griffin by dress, but jerking like a puppet. I get a foot up and slap him away with it. Then there's a monk. A woman, young. Surely no threat. I'm very engaged in wrestling the dog. When she puts an assault rifle to my ribs I understand my error.

There is a flash of steel and the gun isn't there any more, nor the arm holding it. I have a brief image of the boy, Malkin, standing over the downed monk woman. Staring at her. The first time he's drawn blood in earnest maybe. His face set and terrible.

The Factory dog takes advantage of my distraction and throws me off. Stamps on my chest hard enough to make me see stars and feel the metal cage about my ribs creak. Turns on the boy.

Malkin has a moment to flee, and doesn't. Because they've told the kid about what a warrior does, and somehow fleeing to fight another day never figures in those stories. Malkin's face is dead white, like a corpse's already. He lunges forwards with his blade. Buries it in the dog. Leaves it there and comes up with the monk's dropped rifle.

The dog bites down on it, crushing the gun in the vice of its jaws. Malkin falls back, all fingers present but only just.

I get between them. I grab one of the dog's arms and haul it back, strength against strength. I get my jaws in where the mass of its round shoulder meets its thick neck. I barely break its skin. My sheer weight does more to throw it off its strike.

There is a ferocious convulsion that whips through the muscle of it, like an electric shock. I think it's the jaws going into the boy, but it's Leon's blade going into the dog. Driven up to the quillons – is that the word? The blade he meant for me.

The dog is falling. I collapse across it, feeling almost as if he did actually stab me after all. Every joint of me is stressed, the natural and the artificial. I am a comms engineer. What am I even *doing* with my life?

Leon drags his sword out of the beast and his knee explodes, the bullet passing straight through, shattering the joint almost incidentally on its way to somewhere else. No idea even who's shooting. He drops, not screaming but clenched in on himself in pain. Malkin goes down beside him, shouting fit for both of them, panicking. I can help, I go to help.

The call comes in. The plan's gone wrong. *Irae*.

60

ADA

Factory-Bees could not exult. Any such emotional depth had been stripped from her, in her struggles to survive the collapse of the human infrastructure she had relied upon. Ada, trapped in this stripped-down and razor-sharp dream of Bees, felt almost forced to experience the absent emotions on her behalf.

The Griffins without had greater numbers and overall better weapons. Bees had a defensive position, and the Bioforms for when the attackers got too close. More than that, though, Bees was one. Its hacked soldiers were not swift or accurate, puppeteered as they were, but they were a part of Bees. The individual personalities within their heads were cut off from themselves in a nightmare of alienation. They moved and fought as one. And Ada could watch with horror – tinged with awe – as Bees grew more and more adept at wielding them. A swarm of hijacked bodies becoming a surrogate for the insect bodymass she had lost.

And the others, the Factory workers, were at one remove from the swarm. Swifter, able to take individual decisions, working to the enforced loyalty of their implanted hierarchies. Better soldiers than the hacked slaves, for sure, but Ada could almost sense Bees' dissatisfaction with the way they were *apart*. The slight deviations from Bees' wishes brought on by individualistic thought. Not a tyrannical desire to control, as some human autocrat might have had. Because that human's desire to rule would derive from the knowledge that there were other thinking beings who had chosen you – or who had no choice about your control. The twisted shadow of Bees that had come from the Factory did not wish to dominate, nor took joy in subjugation. She simply found the existence of minds outside herself inconvenient and messy, and intended to remedy the situation until there was only one mind. One mind, having one desire: to persist. No discoveries, no exploration, no higher thought, just Bees existing forever. The triumph of the biological Von Neumann Machine.

She tried to feel pity for Factory-Bees, but there was nothing there to feel pity for. As well feel pity for an industrial process. And her fear of it was like the fear generated by a natural disaster. You couldn't hate it, even, or at least any such hate would go slanting away off the imperturbable face of the entity, which could never hate you back or even properly understand that you existed. Save as a slight inefficiency in its procedures.

Ada was that slight inefficiency. A necessary evil, because coring out the individual from the midst of her implants would have compromised their functioning. Because the neural implants were inextricably meshed with her, Bees brought her along with its decisions and commands, showing them to her as though she read the book of them over the entity's shoulder. And because the currents and flash of them passed through channels linked to her brain, she felt. Her emotions poured into the vacuum of Bees' own absences. She feared, she hated, she exalted.

And waited. Because Bees did not even need to win the fight. All the defenders had to do was hold, a position they were eminently well-placed to do. And the attackers had not thrown their all into one monstrous, costly assault on the Apiary walls, probably because plenty of the defenders wore the faces of their friends. The Griffins feinted and retreated, danced about, tried to bait the defenders out. When they got too close Bees spent some blood and bodies chasing them away, but neither side would grasp the nettle.

Attack! Ada tried to tell them. She tried to open a channel to Wells, even as Bees registered the Martian Bioform ranging about the edges of the conflict. Bees had locked away her comms though, becoming ever more familiar with the workings of her implants. Perhaps soon enough she'd find herself imprisoned, stripped of senses, in the dark of her own brain.

There was a new brood of the wasps incubating even now within the walls of the monastery. They had been intended to hack the remaining prisoners currently locked up in the cellars of the monastery, but they would do just as well in a battlefield deployment. And then the numbers disparity would flip decisively the other way and the fight would be over. Ada could only hope that Wells would run then. The wasps were calibrated to human neurology. The Bioform would be immune, but not proof against bullets.

Her comms receivers crackled and hissed for a moment. Internally, she scrabbled for control of them, trying to fight for an open channel. *Wells! Wells!* And she had it, in that second. Some electromagnetic contact had sufficiently startled Bees that Ada managed half a second's panicked pleading to Wells. And even as she tried to make words, she didn't know if she was saying *Attack* or *Run away*. And even then she didn't know what she meant. The horrifying glory of Bees moved through her like the spirit of a god. Bees had made Ada her disciple, her prophet, the bringer of her caustic message. There was a privilege in that, a significance. She, Ada, had finally achieved something in her life. Or been used to achieve something. And wasn't that almost as good?

In the midst of that desperate attempt to reach Wells and tell the Bioform – what? *Bow down and worship me, for I am the herald of the second end of the world?* – she realised what that EM contact had been. Not a signal, nothing intentional at all, but a brief handshake from a nearby friendly system.

Or that had been friendly, once. Another Martian implant. *Irae.*

She felt Bees register and understand, just a fraction of a second later. Her body bunched to flee. She felt the wrench in every joint as the misaligned endoskeleton fought against Earth's gravity and against her own flesh. The pain was like lightning and burning all over and she screamed, and then Bees would not let her scream because it was inefficient. She felt herself about to black out, but Bees refused her even that mercy. She lunged across the small, central room – Prior Stick's office as had been – feeling like hooks were prying every part of her into pieces.

Trying to get away. Because all Bees' attention – hence hers – had been on the attacks outside. She hadn't even considered the idea of a single intruder snaking its way within the walls. There was that disruptive individual thought again!

Irae blurred out of the grey of the walls like a nightmare. Like some alien thing materialising out of thin air, save it was just its camouflage not keeping up with the speed of its strike. It must be dialled up to eleven right now. It must be running full goblin mode.

I'm going to die, Ada understood, because Irae going full heat wouldn't have any room to remember its old friends, surely. It would just be the serpent, the devourer of the sun that she had become.

Irae seized her. The sudden shock of impact made every joint scream. She registered the thin, high whining noise in her ears as the only complaint Bees' iron fist would let her make. The Bioform's jaws had her where the neck met the shoulder. The creature's body had looped about hers. She tried to fight, but Irae was so much stronger and her control of her own limbs was fitful and agonised. With a convulsive sidewinding heave, Irae had her out of Stick's office and down the corridor, moving by a weird combination of limbs and rippling body, Ada's legs trailing loosely behind.

Ada's channel: *Irae.* In that moment, as Bees scrabbled to reorder her forces, Ada had a moment's grace to speak over comms.

Irae's channel: *Sorry sorry sorry gonna hurt you more than me just hold out just hold out just hold out I remember I remember I remember the plan the plan the plan.*

The serpent wrestling itself into a knot, no past, no future, just an eternal present where it coiled itself tight about that single idea. Winging it so close to the sun that every binding principle of its identity was running like wax, but still clutching to that single idea. Irae, on the very edge of the cognitive abyss, with only flames below.

Ada's channel: *Irae, kill me. You have to kill me.*

There were bodies ahead. Bees pulling every free agent back to get in Irae's way, to recover control of Ada, the hub, the beacon of command. The first few were clumsy hacked zombies, not even armed. Ada felt every jolt as the Bioform ploughed past them, over them. Brainless husks of monks bowled over, flailing, uncoordinated. Irae ploughed down some stairs, practically bounding from wall to wall rather than the steps themselves.

Irae's channel: *Got plans got plans got plans for you yes Wells' plans Bees' plans you don't get a say.*

Ada's channel: *Irae, I am Bees.* Feeling the grandeur of it, feeling the fear of it. Weeping, inside. Weeping at the triumph of it. *I am the swarm, the coming world. Bow down and worship what I will become.*

Irae's channel: *Screw you I'm supposed to be the crazy one.*

At the bottom of the stairs was a Griffin with a gun. Ada tried to warn Irae but couldn't complete the circuit of the impulse. The bowled-over monks behind her were the targeting system. The Griffin below lined up the shot based on where Irae's head would be, on a best fit statistical model. The shot would probably blow Ada's shoulder apart after passing through Irae's skull. All of Bees' focus was on the single action, inhabiting body, arm, trigger finger. A level of fine control she hadn't ever tried with the zombies before.

Ada felt the tide of Bees' control pass through her, from the Factory transmitter, via her implants, to the world. But it ebbed from her own body, leaving her custody of her useless limbs. Somehow she had become a part of Bees enough that Bees had written her out of the calculations.

As Irae hit the bottom of the stairs Ada threw her body left. A spasm as undirected as an epileptic fit, but it threw Irae off its stride, sent them tumbling out of their determined progress, end over end. The searing pain of the movement was like hot irons in her. She felt one strut of the exoskeleton tear wetly from some part of her hip, blood abruptly slicking her skin and soaking the inside of her suit. She screamed, full-throat screamed, because Bees wasn't paying enough attention to stop her.

The bullet danced past Irae's head, a breath against

its scales, and then Irae cannoned into the gunman and flattened him, before flailing off down a corridor, tail whipping behind.

Ada felt herself locked out of her body with an almost physical click. She had played her sole card. And Bees was not even angry at the betrayal because Bees had no anger in her.

What Bees had was breadth of thought. While Irae had been descending the stairs, other servants of the Factory had been closing the noose. There were more bodies ahead of Irae. Humans, Factory-worker and slaved, plus some of the Bioforms.

Irae's channel: *No no no no no.* The sense of the reptile-form desperately trying to get somewhere that was being swiftly cut off. A sudden, wrenching change of direction, flailing down a new corridor, then another turn. In the heart of the hive, searching for a way out.

A knot of people, ahead. Griffins with their hacking swords out. Irae rammed the pack of them, scattered them. Clutching human hands slid from its scales, snagged in its temperature control vest. Irae clawed frantically at them. Raked faces that didn't care, carved skin that didn't feel. More hands grabbed Ada and tried to pull her away, twisting every abused joint. Irae's jaws clamped tighter, refusing to surrender her. Irae broke out of the melee, still dragging its prize, its hide gashed bright with blood in two places and the vest gone. Irae'd shrugged out of it. The garment that let Irae regulate its temperature.

Earth was hot, but the corridors of the monastery were cool. Irae's fire began guttering.

Irae's channel: *Got to got to got to – what was I? – yes, got to—*

The big Factory dog slammed into both of them. Ada felt a rib crack, tried to scream again and wasn't allowed.

The dog had a massive hand on Irae's left leg, another at the back of its skull. It lifted it high, pulled, strained, trying to tear it loose, or just tear it apart.

Ada's channel: *Irae, kill me!*

But it wouldn't even help now. This wasn't some hacked body that would fall out of Bees' control if Ada's light winked out. The dog lunged forwards, its huge jaws closing on Irae's twisting body. Ada dropped out of the reptile-form's clutches at last, collapsing on the ground like a heap of sticks. Staring upwards as the serpent whipped and thrashed in the jaws of the hound.

61

DEACON

My mind is full of orders. *Kill the snake!* And that is good. Snakes are for killing. I know that in my head and I know that in the other place, that says *good* and *bad* even without anyone telling me. It twists in my jaws and I want to shake it, just lash it about like a whip.

But Factory Admin's channel also says: *Must not hurt the woman.* And she's screaming. Even the slightest move. She screams and the snake writhes and twists and tears at me with its own claws. I am trying to close my own jaws tighter, my teeth carving into its scales, its blood filling my mouth. But also, because of the way they made me, I cannot breathe properly without my mouth open, and so I strangle myself even as the snake's tail whips about my neck to help it along.

The woman's arm smacks me across the nose, just a flailing motion. It feels like *Bad Dog*. And the Admin channel is telling me *Kill, kill, save, save!*

There is fighting all around us. My ears are full of guns, my nose of blood-scent. I don't understand why we are fighting. I don't understand why we came to this place or what happened to all the people here when we did. Factory Admin said it was a good thing and that they are our friends now, even though the bitter acid smell of what happened is bad in my nose. Everything in me says *bad* except for Factory's voice, but when Factory Admin says a thing then that thing is true. I know this, like breathing. But they made me badly so sometimes breathing is hard.

The snake gets a clawed foot in the corner of my mouth, hooking at me, dragging my head sideways. Its scales are shedding, slipping through my teeth. I feel the muscular flex of it, almost lose it. It skids from my mouth but I snap down again, get a better grip. One blazing eye is right up against mine, pupil a dagger slit. My other eye is staring the woman in the face. The small pale woman, like the small pale man I killed when Factory said. Killed and left out in the world before I met with the little monk and his cart of pups. Death and new life.

The woman's face is strange. They made her badly too, I think. Strange, but I can read it, eyes weeping with pain, screaming, and trying to free herself, and beating at me, or at the snake, or both of us. Factory Admin's voice comes from her to me, and yet her own voice is underneath it, and sometimes I hear it too.

She is begging me not to hurt the snake. She says the snake is her friend. But the snake is my enemy and the snake attacked her.

I don't understand so all I can do is close my jaws harder so that the snake flails and tears at my mouth so my own blood mixes with the snake's. Then the other dog is there.

Not a dog, to the eye. A dog to the nose, though. No dog ever failed to know another. A dog that is to me what the little woman is to a human. She bursts into this passageway all legs and hands, spilling two of our new monk friends in with it. She doesn't even slow for them, moving faster than my weak eyes can follow so that I know her with my skin and ears and nose as she rams into us all. Controlled, somehow. Precise like a machine, like I always wanted to be and couldn't.

She hits me around the snake and the woman. Half the breath goes out of me and I want to let go of the snake so I can meet this new challenge but Factory Admin says: *No! Kill! Save!* Commands cut down to single words because of all the other admin that Factory is doing right now. Almost no time to spare for me. So I fight back, clawing, clubbing, my mouth full of snake and the snake's coils full of screaming woman. I am heavier, I am stronger. She, the new dog, is faster, but she won't hurt the snake or the woman. We wrestle, we fight, ramming back and forth between the walls. She digs fingers at my eyes. She gets a claw under the Factory eye in my forehead and I know a searing pain when it comes away, tears from me. A blindness in a sense I didn't even know I had. Somewhere in my head Factory is screaming at me, *Bad Dog, Bad Dog!*

The snake twists from my jaws but I get it in one hand, right behind the head, and slam it to the stone floor. The woman, half trapped underneath the loops of it, shrieks like a saw in my head. I slap the new dog with my other hand and then get my teeth into her, and hers into me, turning in circle after circle, just two dogs fighting over a snake. Her blood, my blood, the snake's, the woman's. Like we could go on forever and forever.

Then I am not alone. Bellman has come to help, to save me. Bellman, with the Factory seeing out of his dead eye. Bellman with a pistol aimed, trying to get at the new dog, only we're going round and round, neither of us letting go.

Then the snake is out from between us, still coiled about the woman. Its body is loops but its limbs scrabble it away. Bellman changes his aim, desperate. He can't hit the woman. He can't let the snake escape. I see his eye wide, his teeth bared, like he's arguing with Factory in his head. The new dog tears at me and I try to tear back but I need to see.

Someone is there, with a blade. Going for Bellman. Going to kill my friend Bellman. I try to get in the way but new dog won't let me.

My channel: *Bellman! Look!*

The warning just in time. Bellman jerking round, bringing the pistol towards the little human – a child, no more than a child but she has a blade. I see it flash, already in reach. I can't get in the way. I throw the new dog off me with a new strength. I'm going to I'm going to –

The blade carves. Bellman's gun falls away and two fingers. He falls back, clutching his maimed hand, crying out in shock. I go for the child but the new dog is grappling at me. And then the blade is at Bellman's neck.

Factory Admin's channel: *Kill.*

I know I must kill. Doing what Factory wants is always the greatest good, even when it's killing defenceless little chalk-skinned men who did nothing but help. It's always good. It's iron. You cannot not do what Factory says.

I bunch. Bellman is very still, his complete hand gripping tight at his other wrist, as pale with pain as the chalk man was. His living eye on me.

Factory Admin's channel: *Kill.*

The child with the sword is very still, eyes locked with me. A challenge, always a challenge. Dominance. And Factory is dominant. You cannot stare down Factory. I bunch again. I will throw off the new dog. I will kill the child. I will kill the snake. I will save the woman. I am the strongest dog in the Factory. But Bellman will die.

My best friend. Bellman. And Factory is in my head giving orders like it has been from the moment there was an *I* who understood what Factory was. Orders to kill, orders to protect, like a set of bones under my skin that cause me pain and never fit right. So that it was all the other things in between the bones that made up actual life. The work, the eating, the friends, like Bellman.

I feel something break in my head. Something hard and sharp but – I now find – brittle. The pitiless gaze of the child. The strangling clench of the new dog. The hideous taste and look of the snake. Even the lash of Factory's displeasure. They are nothing to Bellman being dead. I hear my own voice, rough and unformed, saying, "No no no no no." No to Factory. No to the child. No to Bellman dying. I drop to all fours. I duck my head, peering up desperately at the child. If I had a tail I would tuck it between my legs.

The snake is gone, the woman with it. Factory lashes me again and again with *Failure! Failure!* But Bellman breathes and the child does not advance the knife. Only then do I understand that half the fear in the room that is not mine is hers.

62

IRAE

I was running so hot back there, no room for thoughts. A mad fight, like something out of a dream. Except I carry enough souvenirs out of it I can't exactly dismiss it. And now I'm cooling, vest gone, any ability to control myself gone. Need a fire, a sun-warmed rock. The unnaturally hot outside of this cursed planet. But got a job to do. It all comes back to me.

I'm slowing. The dial falling helplessly down. Cold because of the stone around me. Cold because of lost blood. Enough energy left in me to flail aside a couple of zombies and knock the guns out of their hands. My own gun is out, but I can't muster enough concentration to aim back and move forwards. I just start shooting wildly behind me. The Factory people – the more dangerous threat – have sufficient self-preservation instincts to duck back. I kill a couple of the others, I think, because they just stand there. I don't feel bad. I'm running cold enough by now that I don't really feel much at all. There's a band of temperature where empathy can survive without denaturing. Too hot and I'm racing so fast that I don't have time to think about others. Too cold and it all seems so distant. So far away.

I lump myself into a courtyard, open to the sky. A little sunlight, a little warmth. But nobody's giving me an hour's basking time to get my core temperature up. I remain this cold thing, and there are more dogs on the way. Not Wells, but Factory dogs, idiot puppets of this enemy Bees.

Ada is still alive. I'm actually a little surprised. She hasn't even passed out, but maybe the thing in her head won't let her. She's stopped screaming – the echo in my head suggests there was a lot of that while the dog mauled me. She's giving out a kind of broken keening sound that's worse. And I hurt. Cold doesn't mean numb. The dog got his fucking *teeth* in me good. Tore me up about my neck and ribs. Lost blood, and if I hadn't had some advanced vascular mods I'd probably not be able to get this far.

This is as far as I'm going though. I collapse, Ada spilling out of my grip. My belly tells me of the thundering footfalls of the dogs.

Ada lies like a broken puppet. I see her trying to move her limbs – at her command, at Bees? – but the endoskeleton is fighting her like a whole third player who won't cooperate with anybody.

Then the warm sun is blotted out. A pair of the dogs are there, growling, snapping. I hiss at them, mouth wide, fangs flashing. The last defiance of a broken snake. The air is full of...

I think ash first, for some reason. Nothing's burning, though. Honestly I'd welcome the fire, just to shift the dial a little, warm me enough to fight one last time. Earth has been a revelation, frankly. Can't recommend it. Bad review for future tourists.

Not ash, though, but bees. And for a moment I think that's just more of the same. So much bad news queued to read that I'm honestly not going to get round to all of it. But no.

Bees. Bees and bees, coming out of every crack and between every stone. A whole fugitive hive-in-waiting here. Waiting for me. Or, rather, for Ada.

They land on her writhing form. At first just a handful, black dots on her clothes, her skin, in her hair. Then more and more. All the surviving bees of the monastery hives, which means hundreds, thousands. The hives are gone but the bees remain and suddenly the air is thick with them, and each one is buzzing. Not just in the ears, but in the mind, over comms, on all channels. I writhe as the sound grows, and then turn off my comms entirely because honestly everything hurts enough already.

Ada flails and thrashes ineffectually. The bees rise from her and then descend again, more and more of them. Like some spontaneous generation where you don't even see where they're coming from, but every second there are more.

I hope you're okay, Wells. No comms now, so probably you think I'm dead or something, and doubtless that fills your life with joy, that I'm not around to piss you off any more. And probably I'm on my way out. Running too cold for a proper diagnostic. But I cling on to see this, at least. I see Ada swarming with bees. With Bees, capital B. The Bees we brought, now in control of all those bonus insects. Infiltrating them enough, taking over all those tiny minds. As cruel, on that miniature scale, as the Factory. As brutal in its necessity. Do I feel pity for all the little lives she's stolen? I mean, no, but mostly because I'm out of that narrow span of the dial where pity really happens for me.

Or maybe for them it's an apotheosis. A raising up to the divine.

The bees descend until I can't even see Ada. Just a vaguely humanoid mass of insects that thrashes and kicks. And the dogs are right there but they've fallen down. They also kick and writhe. They clutch at their heads. There is so much going on, on the EM band. If I crack even the least sliver of a channel it's like opening a window onto the surface of the sun. In the midst of those crawling insects are the hive units of Bees, each one transmitting interference.

The voice of the Factory, transmitted through Ada, is gone.

I twist my head. Past the whimpering, whining dogs, I see the zombies. Monks, Griffins. I see them clutching at the backs of their necks, dragging away hunched little things that were nestled there. Casting the curled-up dead-looking critters away or stamping on them. Weeping, crying out, all that useless human stuff. I see Factory humans and Factory dogs twisting, trying to block up their ears even though what saws through their heads isn't sound.

And gone. The release of the EM barrage is like coming out of a high-pressure chamber onto the Martian surface.

The bees ebb back from Ada's body like a bristling chitin tide. Left behind, stranded, she is small and pale and delicate, bloodied, twisted of joint.

She opens her eyes. Forces her arm to move. One hand on my own red-smeared body.

Tentatively I allow channel access again.

Ada's channel: *Irae.*

My channel: *Hurts. You back?*

Ada's channel: *I'm back. The connection's gone. To all of us.*

My channel: *All friends then?*

The sun is blocked again. I roll my head round. The dogs are there again. They do not look like friends. They look lost. They are cut off from the Factory, and only now do I consider that doesn't suddenly make them Good Dogs.

Wells explodes out of the building behind them, bowls them over. I have never, Wells, been so glad to see you. Even cold as I am, I find that I can welcome the sight of your ugly mug.

COLONY COLLAPSE DISORDER

The dogs are fighting, Wells against the others. They are desperate, berserk in their panic. I am no longer fit to be a melee combatant, honestly. Too cold, too torn up. I have my gun, though. It's come through all that and still tilts and pivots about on its mount perfectly well.

My channel: *Wells, get the hell out of the way you idiot.*

Drawing a line on one dog's skull, on its eye, on the pupil of its eye. I'm that good.

Wells' channel: *No, wait.*

My channel: *Seriously?* Even as the dog gets a paw on her, and then the pair of them are bearing her down. *Oh fuck it.* Adjusting my aim, about to send someone to live on a nice farm somewhere.

Bees' channel: *Hold fire.*

Bees is not the boss of me. Nonetheless, I hold.

The two dogs are barking furiously into Wells' face, but they're backing off.

Bees' channel: *We have introduced ourself. We have made a connection. They were always told that, if they found themselves alone, everyone was their enemy.*

My channel: *They're damn right about that.*

Bees' channel: *We are telling them who we are, all of them. That we are another self of their master.*

My channel: *That going to work, is it?*

Bees' channel: *We shall see. We have an understanding of the channels and pathways our opposite number used. We need to disable their hierarchies.*

My channel: *Or you could keep them. Just switch them to you.*

Bees' channel: (A very considered pause.) *Yes. We could. That is a thing which is possible.*

I feel judged.

Wells drops down beside me, beside Ada. Gently lifts

Ada up into a sitting position. We've all three of us seen better days. Then Wells has her medical kit out, bandages and painkillers and a bunch of antibiotics and other drugs calibrated to our various biologies. And some heat packs. Bloody marvellous. Each one like an extra notch on the dial. Life and thought and speed in a chemical reaction, better than drugs.

63

SERVAL

It wasn't exactly clear how Pardoe died. Like most of the Senior Fealty – the ones who'd been around for a while, certainly; no old, bold soldiers as the saying went – he'd been towards the back. And the fighting at the Apiary hadn't even been that bloody. The total of actual dead was eleven, plus another four who would probably not make it, and a further nineteen injured in some way or other. The total dead amongst the defenders was seventeen, she'd heard, and several of them had died in the tightly localised action *within* the walls, not skirmishing with the Griffins outside. The whole point of the assault had, after all, been show. A distraction to let the glass lizard monster do its work within the walls.

Pardoe had caught a stray bullet, it seemed. At the back, where only the women might have seen. Elsha confirmed nobody had. Absolutely nobody. Had told Serval so herself. So they were all left just with the knowledge that the man was dead. And if it had, presumably, just been bad luck on his part, well, he would get all the honours of a dead Griffin nonetheless. One of his sons would assume his role, as head of family and head of Pardoe's personal Fealty. And, hopefully, not be a problem for at least a few years as he fought off his own challengers, and quite possibly his brothers too. And maybe Leon or Serval could work on the boy – or an obligingly ambitious brother – to blunt the fangs of any future threat.

Pardoe, by whatever means, was dead, anyway. And Serval had enough self-awareness to know this was no great victory for truth and justice. It was a victory for her and for Leon and for Malkin, that the chief challenger who'd stalked them for years was now out of the way. It was, she suspected, a victory won because the future that she might craft for the Griffin was more palatable to outside interests than the one Pardoe and his old guard might want. Which outside interests very much included Elsha's friends, whoever they were. But Serval was worldly and wise enough to understand that and still take the win. Security today was worth trouble tomorrow most of the time.

Right now she felt anything but secure.

They were in the Den, the throne room with its wheel of swords. All Senior Fealty was assembled – the only gap in their ranks being the late and ostentatiously lamented Pardoe – along with as many Juniors as could fit in the room and still leave a space to fight in. Because there was going to be a fight. It had, after all, been promised. And nothing that had transpired in the interim had removed that hanging sword from over her head.

On the throne, Leon sat. The man who'd originally held the sword. His leg was up and heavily bandaged. Not too much pain, thanks to some of the drugs from the bunker's own pharmacy, and something the pallid Martian woman had given, too. Their landing craft had machines that could just spit out drugs, apparently. A thing of the Old that Serval was already coveting.

Leon gripped the arms of his chair, and probably it did look like pain to the others. Serval knew it was tension, though. The frustration of an old man who's had to pass the baton onto another.

Malkin had chosen a more slender sword from the wheel. The great hacking claymore that Leon had used would be outsized in the boy's hands. He wore a stab vest over leather, but no helm. Serval could look into her son's face and see... fear, bravado, a desperate need to prove himself.

His opponent slouched into the room from the far side. The monster, the dog-thing on its long spidery limbs, its dreadful bat face twisting left and right on its short neck, ears swivelling. It had come out of the fight at the Apiary relatively unscathed, worse luck. Nor had it just slunk off into the world, or back into the heavens. It remembered that the Griffins were owed a fight.

The herald kid blarted his airhorn with undue enthusiasm. The dog creature flinched back from the unholy sound, baring its teeth enough that everyone eddied back from it. Malkin took a deep breath, biting at his lip, and stepped forwards, levelling the sword. The whole assembled Fealty fell silent.

The beast cast about almost blindly before its snout fixed on her boy.

Leon shifted awkwardly in his seat, twisting about the fixed point of his injured leg, up on a stool in front of him. "This fight we were promised," he said hoarsely. "For the bad blood between the Griffin and the world of Mars. For our dead and the false accusations levelled at us. For the honour of the Griffin," he said hoarsely. Invoking the name of the man who suffered and writhed, limbless and undying, in the Reliquary below them.

We are such a twisted thing, Serval thought. *We are what he made us, and he is what we made of him, round and round, the serpent devouring itself.*

Her gaze was so fierce, fixed on her son, that she thought he must feel it like a skewer. But Malkin had grown and stepped from the shadow of his parents and had thoughts only for the fight.

"Fight, son."

Malkin moved. And, even though she knew how much had been talked over beforehand, Serval's heart was still in her throat.

Her son drove down the centre of the circle, blade flashing. The monster seemed to take up all the other space. Its horribly long arms could reach anywhere. Its muzzle split in a bristle of fangs. It seemed to hollow itself around Malkin's stroke, surrounding the boy, a world of darkness to that one light.

It came for him with claws and he rolled aside, striking backwards and nicking the thing's hide. It rounded, snarling, playing the fiend, slathering and snapping like a rabid thing. The onlookers eddied back despite themselves.

A grasping clawed hand came down. Malkin dove aside, kicked back, came up with his blade at the thing's throat, its paw clasped about his waist. Every eye was on them.

The sword twisted, moving the edge across the beast's hide, trimming the hair there. Serval saw the monster take a huge breath.

It let go. It fell back. It ducked down. Malkin shadowed each movement, keeping the blade in position without striking. The beast sank down to its belly, one hair-nested eye wide and on Malkin. Serval could read the tension in it. The moment when it found out if its trust was justified.

Malkin seemed to have a similar moment of indecision. How much more of a hero, if he struck now, lopped the thing's head off, declared war on Mars and robbed them of any chance of drug printers and engineering know-how and all the rest, all the bounty that might flow from that quarter? But Malkin would be a hero. Fealty would cheer him to the echo and nobody would ever doubt his right to take his father's place in time.

Serval could barely breathe, waiting to see who it was her son would become.

He made his decision. A shudder went through him, of other futures let go, to blow away on the breeze. Malkin stepped back.

"Live, monster," he said. Soft-voiced but the loudest thing in the room. "Live, by my word."

An uncertain silence, for just a second. The men of the Griffin recontextualising how things were done, and what a victory looked like. Then someone was clapping, cheering. One of Senior Fealty, even. Old Storri, Elsha's unimaginative and complacent husband, doubtless his own idea, not prompted at all. And once he was at it, more joined, and then all of them. She could see the ripple pass through them, see just who it was who was happiest, who the most reluctant, who to keep an eye on. When Malkin raised the sword, though, the roar of it all redoubled, and she knew he had them, at least for now.

And if we can bring real bounty from these visitors, then who knows... We can call it tribute. The spoils of my son's triumph...

There was a meeting, later. Away from the mass of Fealty, behind closed doors. Leon – sitting up in bed now with his dressing changed, Serval, Elsha and the awkward pale woman, Ada. And Serval felt obscurely proud that her husband knew well enough to stay quiet and let the women sort things out for once.

"I have been told of your situation, regarding... *him*," the Martian woman said in her weird, nasal voice. "There are ways... In the long term it might be possible to bring a facility for genetic editing from Mars. It's something we are very reliant on, so we've become extremely good at it. The markers in your genome that make you dependent on Griffin's formulae, they can be removed. Your children's children, at least, could be free of it." She bit her lip, sagging lopsidedly in her chair. "I appreciate... it's a thing. A tradition. But still..."

Serval had already thought ahead. A ritual, the passing round of cups. What needed to change? Save that going Without would no longer be a sentence to pain and death, just a social slight. Her grandchildren, everyone's grandchildren, no longer being shackled to the bunker and its grisly occupant. No worries of supply as the population grew. She was already nodding before Leon did.

"In the short term, we can isolate the actual substance, print out a supply. There would be no need to..." And Ada wanted them to put Griffin out of his mad misery, of course. But until she could free them entirely of the man's chains, then the bunker's founder would continue to suffer. Nobody was going to go from being dependent on the Old One to being dependent on their new Martian friends. That was the sort of pressure a friendship couldn't survive.

"And there are some other things," Ada said. "Your systems here. Some of it is fixable. Heating, ventilation, drainage, even communications. The infrastructure is complete. Some of it can even be told to fix itself, it's just you've lost the ability to talk to it. We can restore that."

"Why?" Leon grunted.

Ada closed her eyes. "We came to help," she said. "Tecumo came to help. To help the Factory, as it happened, and... that didn't go well. But we came to help and we can help you." Her eyes flicked open, angry suddenly. Just a small, pallid creature that could barely prop herself up in a chair, but Leon flinched away from the fire within her.

"You're not who we were looking for," Ada spat. "You are... not the sort of people we have, on Mars." The specific accusations hovered on her lips, struggling against diplomacy to be said. *Violent bullies, thugs, brigands.* Diplomacy won out, though, and she said, "But you could be worse. And perhaps you will be better. The Crash – the end of the Old – it hit hard everywhere. It turned Bees into the Factory. It almost extinguished us on Mars. It enabled men like Griffin to gather people like your ancestors, and let him do what he did to them. We've all suffered. And now we can help you. And you can help each other. You, the monks, all those villages. Help, not raid or rule or fight stupid little wars. Help and rebuild and live."

It all sounded profoundly optimistic, and spoken by a woman with only a passing understanding of how things were, here in the Arreno River region, or on Earth as a whole. But Serval would take optimism. Perhaps it would even be justified.

"And the Factory?" Leon asked. "My people want to take it."

"Under new management," Ada said with a sickly smile. "The Bees we brought has – I don't know, merged with? – anyway, taken over from the Bees that was there. Brought her personality and priorities and installed them in the Factory processes. Trying to help. Trying to improve things there. Build a better dog, no hierarchies. And she's at the Apiary too, now. For real."

"And if we don't behave we get stung," Serval finished with a hard smile.

"That is between you and Bees," Ada said. "But when I'm gone, it will be Bees you can talk to, about tech things, things from the Old. I know you had a kind of situation with the Factory, where you basically left each other alone. Because Bees was stuck in there, and because you'd tried and failed to attack the place a while back. So you could go back to doing that, I suppose. Or you could open a channel, start talking. Bees has people there who she's responsible for, and who have lived all their lives with a leash in their heads, that demanded unquestioning loyalty. The villages will be talking to them now, I'm sure. Working out what they can get out of the new regime. Do you want to be left out?"

That much talking seemed to exhaust her. Serval looked to Elsha. The woman was nodding, seemingly only half paying attention. Some part of her mind elsewhere.

64

THE WITCH

I'm in the room, of course. Or Elsha-as-I is, which is the same thing. Because I don't have fancy Martian implants to piggyback off, my nomadic body is having to hunker close enough to the Griffin compound to establish a link. This is my life, crouching on hillsides and in dells, in all weather. Wandering the wilds, gathering mushrooms and occasionally being run out of town by suspicious villagers. Nothing anyone would want to emulate, or so you'd think.

"You'll appreciate," I tell Orme – even as Elsha listens to Ada inside – "that I'm a little wary of the proposition."

She sniffs, a little of her old sharp temper. "Well of course you are. But believe it or not, it's sincere. If you can do it. You'll understand I'm unsure of the process you use. Whether it needs to be virgin maidens or the dark of the moon or some particular genetic lineage."

"It used to be," I tell her. "Right at the start we were actual clones, genetically identical, to achieve the synchronicity of neural structure that allowed a match and a connection. Over the years, though, I – the wider I – worked on improving the procedure. Cloning facilities were, after all, difficult things to keep going when people like *you* were trying to exterminate me."

"Like me, but not me," Orme said. Then, waving away

my sceptical look, "I'm not saying I wouldn't have, if I'd had a reason, but you were never really on my radar."

"Anyway, I managed to generalise the process, hone the implants. The current system involves an invasive fungal graft that basically creates its own supplemental neural structure. It's not as powerful as the original setup but it's independent of outside help and it adapts to any neurology it encounters. Creates its own interneural compatibility. Hence yes, it's possible."

Things are going to be changing around here. Hopefully for the better. I won't say that the grand alliance against the Factory will somehow magically give rise to everyone dancing in a circle singing happy songs, but it's something. That plus the injection of repair work and new tech the Martians can bring us might win us some stability, reset the Griffins from brigand-kings to something a bit more palatable, build a future where wandering old women aren't turned out of doors quite so often…

Or start a new world war. I've seen sufficient reversals in my long life that I shouldn't let myself get carried away with notions like 'hope'. I will always have a retreat ready, an obscurity to scuttle into, where I can preserve myself until the next time.

But right now, I'm in a position to get my hand on the tiller of this new course. And I will modestly suggest that most futures where things are better for me, are better futures for just about anyone else except autocratic assholes who want to control everything.

And perhaps that means I do need to make some personal changes.

"It's just," says Orme, "seeing Griffin made me wonder… when is it my turn to reap? I feel like I've been sowing for years. Since the fall of the Old. Just living my life for the sake of living my life, and never reaping anything at all. Barren fields year after year, generation after generation, Building nothing, just *being*. And old Josh Griffin, he's *being*, too. In the worst way, but he's still around. And I looked at that and thought that he and I, we'd basically contributed about the same to the world. I mean, fuck." More dispirited than I've ever seen her.

"You don't think I'd be an asset, as a part of your collective?" she demands, suddenly combative again. "Having a part of you that can just go on. I mean, who knows about *forever*, but certainly for multiple human lifespans. I've got centuries in me, still. You can't use that kind of continuity?"

I could, certainly. "You think, perhaps, you'll take over. Become the controlling shareholder, become *me* in a way that *I* become *you*?" I ask. "Is that it?"

She manages a brittle little laugh. "It's like the old joke, isn't it? I wouldn't want to become member of any club that would have *me* as a chairperson. I just... I'm tired of being just me. I've lived for just me for two centuries and change. It doesn't make anything better and it doesn't even make me happy. It just makes me sick of being me. Maybe that's the cure for sociopaths? Be one for a few lifetimes and you realise it's not actually the winning hand you think it is. That not betraying your fellow prisoners actually works better, long term."

"Well," I say, aware that for me this is as big a decision as working out what to do with the Martians and the Griffin. "It's doable. It's possible. I'll think about it." Knowing that I'm already thinking through the possibilities of having Orme as part of myself, her attitude, her longevity, and that I'm going to say yes.

65

WELLS

Irae's channel: *This is ridiculous. You're not doing it.*
My channel: *It's my decision.*
Irae's channel: *Then you're not fit to make decisions. Ada, tell her.*
Ada's channel: ...

Ada looks at me miserably. And she could have had herself shot full of sufficient painkillers to take away the misery, but then she wouldn't be fit to make decisions herself. And so the weight of Earth and the jigs and jags of her misaligned implants still pain her, the endoskeleton she's crucified on, moment to moment. All the extra torn tissues and bruises she got when Irae made off with her. When we saved her.

But it's finite, now. After the considerable discomfort of take-off, she'll be up in zero-G, in orbit. And then in flight, between worlds. Long enough that, by the time she arrives back at Mars, she'll be well. She can divest herself of everything they added, that failed to make her Earthworthy. She can, I suspect, swear never to leave Mars again.

By the time she arrives – radio-waves flying so much more swiftly than ships – the new expedition will be ready to depart too, if it hasn't already left. A new crew, a wiser acclimatisation procedure, supplies. Pharmaceutical printers, not the least of it. A grab bag of things that were all invented on this benighted world, and now need to be carefully reintroduced, as though we were reconstructing an ecology. Probably the first of many expeditions. Not a Martian invasion fleet, but at least cordial relations, that we can spread to other parts of Earth still holding out. What will we become in their eyes? The monstrous people of Mars, peddling fantastical technology and benevolent philosophy. Monsters, djinn, devils, saviours. A whole new mythology for the people of Earth to construct. Or else will we find blind hostility that refuses our good intentions and our gifts? Or enclaves where technology has been maintained, and even improved. Perhaps we'll find some domed city where we will be the barbarians begging for table scraps of their knowledge. It's a big planet. Bigger than Mars.

I intend to at least put in some ground work.

I am not going back.

Irae is apoplectic. Perhaps because it reserves for itself the role of trickster god in this pantheon. If there's trouble to be made, it's Irae who should be making it, but here I am flatly refusing to go home.

Irae's channel: *I thought this place was driving you mad getting up your nose and in your ears.*

And it was. When I first came to Earth it was sensory overload, the air so rich with life and sound and scent compared to sterile Mars. And it still is, but my senses have caught up. Adapted. So that I'm not overwhelmed, but it's all still so rich. As though I've been on a starvation diet all my life until I came here.

Irae's channel: *They tamed you, is that it? One pretend fight and suddenly you're that kid's dog?*

I think of Malkin, who probably is a dog person, and probably wouldn't be the worst person to own a pet. We fought alongside one another, in the Apiary. Which is the sort of nonsense the Griffins would trot out, as a reason to do something. I am not some Martian noble savage out of Rice-Burroughs, though. I am not here to play sidekick to some human hero.

I am, in fact, a comms engineer. And while Ada and Irae fly the millions of kilometres back to Mars, I can work on establishing a proper radio transmitter here, to see who else on Earth feels like talking. Give me a year and a workshop and I can probably get a satellite or two into orbit, to relay signals to the far side of the planet. On Mars I am one of a whole class of the technically educated. On Earth, I am a rare repository of knowledge, and they need me, and we came to help.

Although I would be lying if I said that the savagery didn't appeal to me, even just a little. Lying if I said I didn't like Malkin, who is flawed and selfish and greedy and brought up to idolise acts of violence. And who could be so much worse. It is not going to be One Boy and His Dog and it's not going to be The Dog and Her Boy, but I think we are going to get on just fine.

And then there's Cricket. The little monk, the wet one, the useless one. Except he got out of the Factory and told us what was going on and the truth about Tecumo – meaning that I and the Griffins could actually put our grievances down long enough to build this fragile rapport. And then, faced with a lot of people really quite keen for a fight – including me – he stood up to talk us down. To speak for the voiceless, meaning those from the Factory. Which makes him a good person in a way Malkin isn't, and probably I'm not either. Cricket is talking about neighbouring communities outside the Arreno River region. Because the Apiary has contacts made by mendicant monks over the decades, and now it really does have a mandate from Bees. I see a time when it's Malkin on the Griffin throne and Cricket behind the

Prior's desk, and HumOS still being herself and expanding her far-spaced mind across the land, and me... The Martian ambassador, the alien dog-monster to give them nightmares, yes, but to show them how much grander the reach of life is than their little backyards. We will bring order and we will bring wisdom, and we will teach them to dream.

66

CRICKET

After the dry words crackling out of the ether, Cricket felt he had to sit down. And junior monks didn't generally just lounge about in Prior Stick's office but the old man indicated a chair, and he took it.

"I can't, though," he said. He was barely an adult. He hadn't even been on the Earth two decades. "Siblen Boatman, maybe. Or... someone. Someone older. Someone who knows more."

Prior Stick leant back in his own chair. There was a bandage about the back of his neck, and the Martians had given him a course of drugs to fend off infection and boost his immune system. All the victims of the cockroach hack had the same precautions. Factory-Bees hadn't been particularly bothered about the long-term risks to her puppets.

"I don't think there *is* anybody who knows more, about recent events," the Prior said. "Through no fault of your own, you have been at the heart of it."

Cricket opened his mouth for more objections, but that other voice, that third voice spoke again. Popping with static as it issued from the radio box on the Prior's desk. The Apiary's old transmitter-receiver unit, carefully maintained over the generations, and yet become almost useless for want of anyone to talk to.

Now it had suddenly come into its own. It had become the mouthpiece of Bees.

"Siblen Cricket," came that precise, sexless voice, "it's our opinion that a younger candidate is better. More neural plasticity. A great deal to learn! It is to be hoped Prior Stick has many years left in him. Before his nominated successor would need to take over."

Prior Cricket. Ridiculous, obviously. Or ridiculous *now*. Give him five years of being Bees' chief disciple, living half out of the Factory. Acting as the mouthpiece of the distributed intelligence, its ambassador to the powers of the world.

"Would I need..." He heard his own voice shake. "Something in my head?"

"It would be convenient," Bees said over the radio. "Initially an earpiece, however. The Factory systems for such things are invasive and geared for levels of control we do not wish to perpetuate at this time. Perhaps in the future."

Cricket had walked all his life in the shadow of Bees. He had brandished the name like a threat, to ward off harm. He had felt, at some level, that a benign and distant entity was watching over him and had an interest in his safety.

He had discovered, mere days ago, that all of it had been a lie, and that there was no Bees. Just a succession of Priors who had perpetuated a convenient fiction, in the hope that it would armour the Apiary against the world.

He had discovered, on the heels of that, that there *was* Bees, but it was something horrible and twisted lurking behind the faceless administration of the Factory. And then there had been the War of the Bees, as that old presence had given way to the new one the Martians had brought. And none of it was the Bees that he had grown up knowing. It had all been something of a religious crisis. And what was he left with now?

He understood, then, that nobody could actually know. He, who knew nothing, was yet as wise as Prior Stick in the matter. Probably even Bees herself could not predict how she would develop and what she would become. But if Cricket was her acolyte and human intercessor then he would have some influence on that future. If he wanted the world of tomorrow to hold something like the fictional Bees he had always believed in – the benevolent protector of knowledge and the weak – he could attempt to bring that tomorrow about. There was nobody else.

For a moment he saw himself as old, the age that Prior Stick was now. Sitting behind this same desk here, or some equivalent at the Factory. Would Steward Malkin of the Griffin – another old man by then – come talk grand matters of politics or exchange envoys with greetings and gifts? Would he be some great pontiff to whom pilgrims travelled seeking the guidance of Bees? Would there be a spaceport outside where Martian traders and embassies landed? Would he listen to the word of Apiary monks who had travelled to all the kingdoms of the Earth using the rediscovered secrets of the Old?

Or it could all be on fire, of course. Prior Cricket, cold and old behind his desk as the barbarians pounded on the door of his office. But if it ended like that, it wouldn't be for want of him trying for something better.

"I would be honoured," he said. "I can try." The Apiary had spent generations trying to preserve the least scraps of the Old. Only now did Cricket consider, a vertiginous shift of perspective, that this moment now, this very fulcrum point, would be the Old to generations yet to come. That if he worked very hard, and if a great many chances fell into place, this might just be the first day of the new age of the Earth.

ACKNOWLEDGEMENTS

My usual round of acknowledgements, for everyone who's helped bring this book into being: Simon and Oliver at the agency, Nicolas and everyone else at Head of Zeus, my wife Annie, and everyone who has read, enjoyed and spread the word of the previous books to the extent that I started thinking about a new one.

ABOUT THE AUTHOR

ADRIAN TCHAIKOVSKY is a British science fiction and fantasy writer known for a wide variety of work including the Children of Time, Final Architecture, Dogs of War, Tyrant Philosophers and Shadows of the Apt series, as well as standalone books such as *Elder Race*, *Doors of Eden*, *Spiderlight* and many others. His Children of Time series has won the Arthur C. Clarke and BSFA awards, and his other works have won the British Fantasy, British Science Fiction and Sidewise awards.